Fighting General

■ ■ ■

The Public and Private Campaigns of General Sir Walter Walker

By Tom Pocock

Contents

Illustrations

Maps

Drawn by Doug London

Introduction

In January, 1964, I waited on an airstrip at Long Semado in the remote highlands of Borneo for helicopters bringing a British Army patrol out of battle. These few soldiers of the Royal Leicestershire Regiment had been following the trail of a strong force of Indonesian raiders through the jungle. There had been many such forays into East Malaysia but this was the first in which the raiders had been met by British infantry.

Expecting to face odds of perhaps twenty to one, the Leicesters had finally found the enemy's camp and, without hesitation, taken it by assault. The odds were, in fact, about six to one but the British killed seven Indonesians, captured quantities of weapons, equipment and documents and held the camp against threatened counter-attack.

When the young soldiers arrived at Long Semado, they impressed me deeply. For most it had been their first experience of action, yet they appeared relaxed and confident, content that they had achieved their aim. This impression confirmed those already formed during visits to a Gurkha battalion, the Royal Marines and the Royal Malay Regiment elsewhere in Borneo. This was a remarkable campaign both for the skill of the defenders and their cunning use of minimum force. It might well become a classic of the military art.

To meet the commander who had devised and was executing the defence of East Malaysia and Brunei, I reported next morning to what had been the headmistress's study of a girls' school in Brunei Town. I had only recently heard the name of Major-General Walter Colyear Walker and I was as unprepared, as many others had been, for the experience of meeting him.

Unlike some of his contemporary senior officers, Walter Walker could only have been a soldier. Indeed, his smart appearance and decisive manner reminded me of an officer of the old Indian Army, which, indeed, he was. This air of professionalism was combined with perception and charm and, in the hour that followed, he delivered the most impressive briefing of the many that I had had since I began writing about defence nearly twenty years before.

The strategy and tactics Walker described to me confirmed the impression that this was an extraordinary campaign. It reminded me of one of those Hollywood thrillers in which the hero, rapier in hand, fights and defeats enemies rushing him from all directions. It was not until later that I realised that Walker faced not only the Queen's enemies.

This began to emerge when I returned to Borneo eighteen months later. Walker had been succeeded by Major-General George Lea, departing with no apparent recognition other than the award of a second Bar to his Distinguished Service Order. That this could have been a pointed snub seemed possible when, at the end of the campaign about a year later, Lea, who had worthily carried on and developed Walker's policies, was knighted.

I was to meet Walker again during my travels as defence correspondent of the London *Evening Standard*. As he gave me further masterly briefings – first at the NATO headquarters at Fontainebleau in France and later in York, Oslo and Copenhagen – another pattern of conflict began to emerge. This concerned the position of the soldier in society. Rightly, he was the servant of the people, controlled by their representatives, the politicians. But, in times of stress, where did his primary loyalty lie? How much could he resist what he regarded as misguided political pressures? Was he entitled to appeal to the people over the heads of their politicians?

This was one of the dilemmas facing Walker. In taking the stand he did, his survival as a serving soldier in the face of opposition from supposed friends seems as remarkable as that in the face of known enemies. It is the subject of this book.

This, then, is a study of a soldier. If, at times, he emerges as a somewhat stern and ruthless figure it is because I have chosen to concentrate upon his military character at the expense of the warm-hearted friend and family man known to many. Writing a biography of a contemporary figure is never easy and seldom more difficult than when the subject is hedged about with rules and regulations. I have, however, felt able to draw, to some extent, on the many talks, both official and personal, I have had with Walter Walker over the years. He has also introduced me to many friends – and some enemies – and colleagues, including many serving and retired officers of the three Services and of NATO allies, civil servants of several nations, politicians and others.

My thanks are due to them for the time and trouble they have devoted to help me complete a portrait of the subject as a soldier. To name them would be invidious and could be embarrassing for reasons that will become apparent. A particular reason for this is that in Chapter 6 I refer to operations during the Borneo campaign that were given the code-name 'Claret'. While military opinion now seems to hold that something can now be told about these – indeed that it is important that their history should be recorded – there are some in diplomatic circles who feel otherwise.

While in Borneo I regarded all information about 'Claret' as strictly off-the-record, as did other journalists, and, I should record, my sources for this did not include the two Commanders-in-Chief, Admiral Sir Varyl Begg, then Air Chief Marshal Sir John Grandy, nor their Directors of Borneo Operations, General Walker, then General Lea.

Subsequently, my sources have covered a wide military and political spectrum and are unanimous that much of this story can now be told. In speaking frankly about Walker's relations with his military and political contemporaries, these and other sources – including some of those who found themselves in opposition to Walker and his attitudes – are equally united in the conviction that history must also include the course of personal relations.

My thanks are due to Beryl Walker for much kindness and hospitality. She has followed the rocky road of the soldier's wife with courage, dignity and generosity and I am conscious that, in concentrating on her husband's professional career at the expense of his private life, I have done her less than justice.

I am particularly grateful to Charles Wintour, the Editor of the *Evening Standard*, both for giving me the time and encouragement to write this book and for being its originator by sending me to report the war in Borneo.

TOM POCOCK

Chelsea, London, 1973

1

Home to India

As a breed they would soon be extinct. The economic and social forces that had shaped the Walker family – and thousands of British families like them – had also created the British Empire. In the twenty years between the First and Second World Wars it still seemed to most of the British that theirs was an Empire upon which, as they said, the sun would never set.

Such confidence and pride in achievement so near to the edge of climactic change cannot have been seen on such a scale since Imperial Rome. These British families who had built and still ruled their Empire were driven by many motives. There was, of course, the urge to make money and the untapped resources – human as well as animal, vegetable and mineral – of the Empire seemed limitless. There was desire for glory on the field of battle or in exploration, finally offering rewards of rank and title. There was an eagerness to do good, physically or spiritually, and, in medicine and the church, there were opportunities for women as well as the men they bravely followed across distant frontiers.

In the nineteen-twenties, a young Englishman of middle-class upbringing could begin a career in almost any field of his choice in almost any part of the world with an interview in Whitehall or the City of London. The horizons of his future stretched from the Falkland Islands to the Yukon; from South African diamond mines to Burmese teak forests; from Australian sheep stations to Cypriot citrus groves;

from unmapped African jungle to the commercial palaces of Hong Kong. As *Imperial Military Geography*, the textbook for defenders of the Empire published at that time, began, 'Covering one quarter of the land surface and comprising one-quarter of the total population of the globe, the British Empire includes almost every variety of climate, race and production.'

Nowhere made more appeal or offered such a challenge to the British as India. What they would describe as the brightest jewel in the imperial crown was a sub-continent, an empire in itself. The Emperor of the Indian Empire might be the gruff, Anglo-Hanoverian King George V, holding sway over its 1,800,000 square miles and its 353,000,000 people – one-fifth of the population of the world – from Buckingham Palace. But even his British subjects, who ruled directly on his behalf, would say that it would take far more than one lifetime to get to know, let alone understand, India and its people. This sense of the unattainable combined with a deeply mysterious half-knowledge of other civilisations and other religions made India a stimulant to the British. It was the spur to ambition but could offer frustration and disaster as well as success and fortune.

It was to India that the Walker family was bound by ties almost as strong as those with England.

The attitude of the Walkers and their like towards India was practical. They accepted mystery as part of the Indian scene – there was always talk of miracle-working fakirs, bizarre religious customs, like the self-immolation of widows, 'The Indian Rope-trick' and the tantalising stories of a Shangri-la that filtered south from Tibet – but they sought to impose upon India sensible British ideas of democratic government, the rule of law, ways of conducting business, hygiene and even morality. Their families' property, offices, farms, factories or plantations were in India and, while these lay some five thousand miles and a journey of at least six weeks away, yet India seemed nearer to the old family home than the opposite extremity of the British Isles.

These people were called Anglo-Indians – a term which later was applied to Eurasians of mixed British and Indian descent – and their aristocracy were those families which had had roots in both countries since before 1857.

That was the year of the Indian Mutiny. Before the Mutiny, everything had been different, it was said. There had been wars, of course, but this was only to be expected in a sub-continent well-peopled with what were called 'martial races', who, when subdued, would make fine material for a vast and invincible army led by British officers. Although inherited memories of the years before the Mutiny were dim, they were lit not only by gun-flashes but by the glint of gold. They were the years before British women – the 'memsahibs' – had settled with their husbands in India – when barriers of race were lower and a man was free to make a fortune and take his pleasure as he pleased. And, far back, were the early memories of the East India Company, of Clive and Warren Hastings, of great nabobs and incalculable riches.

The Walker family of Tiverton and Assam belonged to this aristocracy. To them, as to the others, India had given pain as well as pleasure. There had been a Dr Adam Walker who had sailed from India in 1849 as chief surgeon to a flotilla of warships bound for Burma, where in the Arakan swamps and what came to be known as the First Burmese War, he died the following year. His son, Thomas Walker, had been a 'colonial engineer' and architect, ending his days in Hong Kong. His son, Thomas, had been a soldier in India and had survived the Mutiny into what was regarded as 'modern British India'. His son, Arthur, who had settled his family at Tiverton in Devon, had been a tea-planter in Assam and served with the Assam Valley Light Horse before going to the wars in South Africa and France.

Arthur Walker had married another of this stock. His father-in-law had been a doctor in the Indian Army Medical Service and had died on the polo field. He had had so many heavy falls from his pony while playing this violent game that his wife had implored him to give it up for the sake of his family. He agreed on condition that he could play one final game. It was his last game for he fell and broke his neck.

Arthur Walker and his wife Dorothea had four sons – Edward, Walter, Peter and Harold – and two daughters – Beatrice and Ruth – when they settled at Tiverton. In their substantial Regency house, India never seemed far away and it was taken for granted that some of the boys would first consider making their lives there, at least one of them in the Indian Army.

There were not only such constant reminders as the thick, illustrated catalogues of the Army and Navy Stores which offered its immense range of accoutrements to the British way of life as easily in Bombay, Calcutta or New Delhi as in Victoria Street, London, and the family reminiscences of P. & O. liners in the Red Sea and at Port Said, Indian hotels and rest-houses, bungalows and servants. There were faded sepia photographs in an album of the boys' grandfather, Colonel Walker, surrounded by little men with darker skins and narrower eyes in high-buttoned uniforms and pillbox caps, who, they were told, were 'the Goorkhas' he had commanded.

The spirit of Colonel Walker sometimes seemed to preside over his descendants. His trim, alert features – half-hidden by beard and spiked moustachios of extraordinary length – looked out at them sharply from the frontispiece of his memoirs *Through the Mutiny: Reminiscences of Thirty Years' Active Service in India, 1854–83.* There his achievements, recounted with a becoming modesty, could be read with pride.

Lieutenant-Colonel Thomas Walker's thirty years in India seemed like a list of battle honours. During the Mutiny, not only had he commanded 'The Forlorn Hope' ladder party at the storming of the Water Bastion at the Siege of Delhi but had fought in ten other named actions.

Forgotten battles not even recorded as street names in Victorian suburbs, they included the assault on the Heights to Sona, the relief of Moradabad and the capture of the forts of Rewarrie, Jujjhur, Kanonde, Kurrunknugger and Bullumghur. Then he had fought with the Bengal Native Infantry and the Bengal European Fusiliers. Later and throughout the Naga War of 1879–80 in Assam he had served with, and then commanded, the 44th Bengal Native Infantry, later to be known as the 8th Gurkhas.

His memoirs of these campaigns demonstrated his belief that an officer and a gentleman should take warfare calmly, as the natural order of things for which he had prepared. One morning during the Siege of Delhi he and a friend spied some approaching enemy through a loophole. The friend 'had a short rifle and shot four men while we were breakfasting', wrote Walker. 'He let me have some shots, and I accounted for three more.'

But his life in India was not as easy as his memoirs made it sound. His wife, Maria, died of cholera at the age of forty-two, while he was commanding his Gurkha regiment in Assam and he sorrowfully enscribed her tombstone with an optimistic text: 'And all wept, and bewailed her: But He said, weep not; She is not dead, but sleepeth.' Such, they comforted, was the price of Empire. Colonel Walker himself died in retirement in Canterbury at the age of sixty-five in 1903 – 'Today Shalt Thou Be With Me In Paradise' – and he left five sons to carry on his tradition.

A battered bugle hung on the wall in the Tiverton house, a British bugle that had been captured by the Boers and recaptured by Sergeant Arthur Walker of Lumsden's Horse, a yeomanry regiment. Arthur Walker liked to tell his sons about the South African War, of his own adventures and how his brother, Walter, had been killed defending the outpost he commanded against hopeless odds. Arthur had also served his country in France during the First World War, first with the infantry then, after a serious illness, commanding a labour battalion, and it was part of the family tradition that, but for his patriotism in so doing, he would have been a man of means.

Had he not volunteered to fight for the Queen in South Africa, Arthur could have stayed in Assam at the family's tea garden, the foundations of its prosperity. The garden was called 'Tara' and Arthur had named his Devonshire house after it as a reminder of grandeur he might have enjoyed instead of trying to bring up a large family by chicken-farming. Arthur Walker had been a notable sportsman. A fine horseman, he had won the Master at Arms trophy three years in succession while serving in the Assam Yeomanry Regiment and was said to have been the best polo-player of his day in Assam. A good shot, his feats against big game were recalled by the number of tiger skins, mounted heads and horns and snakeskins which decorated the house in Tiverton. There, surrounded by silver trophies from his youth, he made his mark as a formidable tennis-player, well up to county standard.

Something of a Micawber, Arthur Walker was remembered by his son Walter as being rather like a Malay: charming, happy-go-lucky and eternally optimistic but hopelessly impractical. Even as a small boy Walter was horrified by his father's casual attitude towards the future. But Arthur Walker was a lovable person, widely recognised in Tiverton when bicycling to watch a cricket or Rugby football match at Blundell's, the local public school, or calling at a public house for a clandestine drink or to put a few shillings on a horse.

His wife Dorothea was in striking contrast. A forceful, handsome woman, she worried continually about the family's finances which she tried to keep in order. She organised not only her children and her house – with one maid – but the Mothers' Union, tennis and hockey clubs and rode a motor-bicycle. While she agreed with her husband that one of the sons should go into the Indian Army to carry on the tradition, the others must try to make some money.

The cost of educating the children was kept as low as possible and all four boys were initially sent to a preparatory boarding school, St Petroc's, at Bude in Cornwall. Walter followed Edward there in 1923 but only stayed for a year. He objected strongly to being taught by women and his reports were so poor that the school asked his parents to send him elsewhere. He was transferred to Norwood School, at Exeter, where the headmaster, a parson, was a martinet but, having played cricket and Association football for the county, was something of a hero in the small boy's eyes.

After his first term at Norwood, the Reverend F. N. Bird wrote of Walter, 'A very low standard of work in most subjects, but is now beginning to work and catch up. Must do his holiday work seriously and realise his future depends entirely on *his own efforts now.'* The headmaster drove the little boy to his books with a cuffing of ears and shouts of 'Idiot boy!' Walter began to work hard out of mingled fear and respect, poring over his books even during the slow, fifteen-mile train journey to and from Exeter each day.

At thirteen, Walter followed his brother to Blundell's which accepted boys from local families at greatly reduced fees. Here he came into his own. At Blundell's, as at every public school, team games were the criterion of a boy's abilities. Cricket and football were believed to develop character as well as muscle and give experience of being led and of leading. Although small for his age, Walter Walker was good at games and, to his father's delight, took to boxing. Arthur Walker had always wanted his boys to 'be able to look after themselves' and Walter even took lessons from the school's boxing instructor in the holidays. His aim was to win each fight with a quick knock-out blow. Many years later the headmaster remembered the final bouts of the school's lightweight championship when 'at the beginning of the second round young Walker leapt from his chair, charged across the ring and knocked his opponent clean out before the latter could realise what had hit him'. And one of the Walker relations, watching Walter on the athletics field, remarked with a degree of accuracy that she was to remember, 'That boy should become a field-marshal.'

This headmaster, Alexander Wallace, was another hero of Walter's. Formerly in the Indian Civil Service and the Indian cavalry, he was ordained while at Blundell's and later became Dean of Exeter. This muscular Christian ran his school 'like a good regiment' Walter was to recall and a lifelong friendship began.

While Walter thought of the headmaster as 'the finest type of Englishman', Wallace admiringly described the boy as 'a tiger', a description that was to linger in the memories of both. When Walter became head of the school's day boys Wallace noted his powers of leadership.

Contemptuously nicknamed 'daybugs' by the boarders, the day boys were a lazy collection of youths before the arrival of Walter Walker. In his own words 'one gave them absolute stick – but we began to be respected'. The day boys became a fearsome pack on the football field and in the uniform of the school's Officer Training Corps their brass flashed brightest of all.

Then Walter was summoned by the headmaster, expecting to be praised again for his zeal. But Wallace said, as the boy remembered, 'You're going too far. You are too strict and too finicky. You are antagonising both the boys and their parents. Of course you must drive them some of the time but if you drive them all the time you will lose your friends. Try to lead instead of drive.' The headmaster stressed this point in the term's report with the words, 'He must not allow his zeal to outrun discretion.'

Walter was quick to learn and, when he left Blundell's at eighteen, grateful parents gave him a silver cigarette case and a scrubbing brush inscribed 'I have earned you more friends than enemies'. And Wallace wrote to the boy's father, 'I regard him as in every way one of the most promising boys I have met and I foresee myself a very distinguished military career in front of him.'

But before this time Walter's future had to be decided. Edward was going to follow his grandfather into the Gurkhas and Walter had hopes of joining the Royal Navy. But his mother's cousin, a serving naval officer, talked pessimistically about the current mass redundancies and suggested the boy go into business.

The Walkers had a contact in the Imperial Tobacco Company at Bristol. So an introduction was arranged. Walter spent a depressing day touring a cigarette factory and visiting the dingy lodgings where he would have to live in Bristol. He returned saying, to his father's delight, that he could never become a businessman and could he instead become a soldier? Edward had just heard that he was too tall by several inches to join a Gurkha regiment which, after their tall British officers had been picked off by snipers at a heavy rate in France and on the frontiers of India, had imposed a height limit of five feet, ten inches. Walter had grown to this height at the age of seventeen and now that his elder brother had opted for the Punjab Regiment he was free to take his place in the Gurkhas if he was chosen.

There was only one vacancy in the 1/8th[1] Gurkhas, which Thomas Walker had commanded, for a British officer during Walter's year of eligibility. He had the necessary family connections with the regiment of his choice but majors, who had joined during wartime expansion, were still being declared redundant. But for his father's enthusiasm, Walter's future chances might have seemed as slender as those in the Navy. A British officer of the 1/8th Gurkhas named Fisher arrived to inspect the boy and did so during a round of golf. Later Walter heard that, if he could get a place in the Royal Military Academy at Sandhurst, and succeed both there and throughout a year with a British regiment in India, he would be accepted.

In the entrance examination for Sandhurst extra marks could be earned for speaking French and it was arranged that Walter stay a month in France with a family whose son would, in turn, spend a month with the Walkers in Tiverton. This was an embarrassing failure. The Walkers would never have considered themselves snobbish yet, like most British families in the upper middle-class – or, as they would put it, 'out of the top drawer' – they observed certain conventions in manner and dress which was considered not only normal for their class but obligatory.

The French family Bruneteau of Rochefort were not out of the top drawer. Expecting a middle-class family, Walter's surprise at the modesty of their house was only exceeded by the sight next morning of his host at breakfast. Monsieur Bruneteau sat at table wearing braces but no collar, cutting pieces from a long loaf with a penknife and gulping red wine with a mouth full of bread. Walter understood enough French to hear Madame Bruneteau remark to her son Robert that he, Walter, was not 'one of them' and was obviously of good family and appeared to be offended by the sight of his father's table manners. The boy replied that it was now too late to teach his father etiquette.

Monsieur Bruneteau was a timber contractor and spent most of his time at work, while his wife did housework and Robert was at his books. Occasionally he took Walter on visits with him – causing acute embarrassment by his habit of clearing his nose into the gutter – but for most of the time the English boy was on his own, rowing on a nearby canal and picking up French from farm labourers and lock-keepers. Eventually he met the local count who suggested a game of tennis and further visits to the *château* gave Walter a sight of French gentility but was not enough to erase the rough accent he had already picked up, which was to shock his examiners while earning marks for fluency and the effort in learning.

Nor was Robert's experience in Tiverton much happier. Undoubtedly an intelligent boy, he earned unspoken displeasure by his arrival in sandals and by the ardour with which he pursued the pretty Beatrice Walker. It was to be nearly forty years before Walter Walker returned to France.

The Royal Military Academy at Sandhurst, which Walter Walker entered as a cadet in the autumn of 1931, would have daunted a less resilient youth. The training was firmly rooted in an all-pervading, over-powering discipline the need for which – so it seemed to the staff – had been clearly illustrated by the experience of trench warfare and infantry assaults during 'The Great War'. Most of the training officers and some of the non-commissioned officers had fought in Flanders and their ideas of land warfare approximated to those of Field-Marshal Earl Haig. The cadets in their charge were therefore trained in the costly tactics which they had been lucky to survive and, as before 1914, stress was laid on the importance of cavalry. While cadets were forbidden to drive motor-cars or ride motor-bicycles, horsemanship was an important part of their curriculum.

Walker was surprised on his first day at Sandhurst when shouted at for failing to wear a hat indoors. Thereafter he accepted the long obstacle course of disciplines as necessary and inevitable. While the new cadets' uniforms were being cut they were required to wear blue suits, which were regularly inspected for specks of dust or dandruff, for Sunday church parades. After six weeks their uniforms were ready and they were encouraged to emulate the most spectacular military figure they had yet discovered: the Adjutant.

Captain Norman Gwatkin was everything that a Guards officer was imagined to be. Had their eyes not been attempting to meet his accusing glare, the cadets could have seen their reflection in his riding boots. He was tall, stiff-backed and so broad-shouldered and slim-waisted that it was whispered that he wore stays. Nothing could be good enough for so fearsome an arbiter of discipline, thought the cadets, who nevertheless admired as much as they feared him.

Walter Walker was relieved to find that Captain Gwatkin was a human being after all, albeit in a traditionally hearty manner. After dinner on one guest night, he happened to look out of a window to see a manhunt in the moonlight. Across the Sandhurst lawns fled the Academy's chaplain with Captain Gwatkin and another officer hallooing in pursuit. They cornered their quarry by some bushes and mercilessly performed the ritual of debagging before returning to the mess triumphantly brandishing the chaplain's trousers. Captain Gwatkin, paragon of military virtues, served to the rank of brigadier before becoming Comptroller of the Royal Household in which capacity his duties included the censorship of the theatre.

Second only to Gwatkin in his capacity to strike awe was a company sergeant-major of the Coldstream Guards who ordained that Walker's company become the champions at drill and in the perfection of their uniform and equipment and who saw to it that they did so.

For a young man with Walker's orderly attitudes there was satisfaction in the simple disciplines of the parade ground. The crash of a hundred brass rifle butt-plates and steel-studded boots on asphalt, the mobile geometry of forming fours and the length of paces required to wheel a long and straight rank of cadets offered the rewards of uncomplicated perfection attained. Moreover, they were told, there was a purpose behind it all. Drill induced immediate reaction to orders and kept mind and body alert.

Riding was compulsory for cavalry were still considered the queen of the battlefield, whereas officers destined for the Royal Tank Corps – and, indeed, for the Royal Air Force – were considered to be socially-inferior mechanics. The parade ground and the riding school combined ludicrously when the cadets set off for physical training in the gymnasium, suitably dressed in pillbox hats, blazers with brass buttons, white trousers and white shoes. They paraded with bicycles, first at ease with front wheel slightly inclined. On the commands 'Prepare to mount! Mount!', they swung their right legs over their saddles and were away, riding to attention, backs straight, shoulders squared, hub-to-hub.

The insistence on perfection in what might have seemed trivialities did not irritate Walker any more than did the daily inspections at shaving parade, drill parade, riding parade and physical-training parade. Indeed, during his first term he was promoted to lance-corporal and eventually to sergeant. But finally he began to question the system. Awe of the adjutant and sergeant-major led to many hours spent burnishing brass, polishing boots and cleaning webbing equipment. Thus not enough time was devoted to the study of tactics, military history and map-reading. Surprisingly to Walker, least attention of all was paid to weapon-training. The rifles, which they oiled and polished so lovingly, they only fired about once a term, then to spend hours purifying them with gun-oil, wax, metal polish and 'four-by-two' cleaning cloth as if they had been defiled by being put to the use for which they had been designed.

To make up for the inadequacies of the training, Walker spent much of his free time at his books, taking his recreation in the Academy's boxing ring, where he sparred with the Army champion, 'Dusty' Miller. This devotion to self-imposed duty was not difficult because, like his friend John Willoughby (later General Sir John Willoughby) the allowance he received from his family amounted to only half a crown a week. While rich cadets destined for smart infantry regiments, or the cavalry, could afford to go racing and take girls to dances, Walker, Willoughby and their impoverished friends saved their shillings for an occasional excursion to London by bus. This might be to watch the annual Rugby football match between Sandhurst and the Royal Military Academy, Woolwich, where future artillery and engineer officers were trained, followed by hearty pints of beer in the White Bear tavern in Piccadilly Circus. It was an exclusively male world for such cadets, yet their zeal was such that most seemed to have little difficulty in sublimating their masculine desires and putting thoughts of courtship as far ahead of them as, say, their eventual promotion to captain.

The young men emerging from Sandhurst to become junior officers in the British Army appeared to conform to a pattern. Each was physically fit, smart of dress and bearing and alert to respond to orders, if not always to take individual initiative. What they tended to lack was sophistication and, indeed, maturity. They were at the age of twenty still, in effect, schoolboys, their personalities still pliable enough to be moulded by whatever military environment was to be theirs.

Social distinctions had already arisen between them. The minority destined for the Indian Army infantry were not likely to come from families of substantial means. A private income was regarded as essential for any young man hoping to make his career in an infantry or cavalry regiment in the British Army. But, in the Indian Army – with the exception of the cavalry – it was expected that an officer would live on his pay and, because of the demands of a career in Asia, this was substantially higher.

However necessary economically, the choice of the Indian Army remained a vocation. An officer in the British Army might well serve in India but, for him, it was only one of a number of overseas theatres in which he might serve and his own future would revolve round regimental headquarters in some garrison town, exercises on Salisbury

Plain and the Edwardian solidity of the War Office building in White-hall.

The Indian Army officer's future was totally involved with the East. To know that nine months' leave could be expected every three years seemed to offer an abiding link with home, yet each officer was given a passage allowance of only £500 to last his entire military career and a first-class return ticket home by P. & O. liner cost about £150. Eventually, if he married, he would have to choose between settling for a life absorbed in and by India, or expensive and tenuous links with home; the children educated at boarding schools and holidays spent with relations – or even at the empty school – when another passage to India was beyond the family means.

Yet, despite the self-imposed exile and the expected restrictions of officers' mess, regimental custom, barracks, married quarters and hill stations, India offered the freedom of wide, almost limitless horizons. Middle-aged and elderly men who returned to the British Isles from a lifetime in India were regarded as having survived a long ordeal. Behind each sun-dried face must be the memories of experience far beyond the expectation of home-bound mortals. They themselves had lived in the vast, violent, mysterious East that, at home, could only be tasted fleetingly in the writing of Rudyard Kipling and P. C. Wren and the painting of Frank Brangwyn and William Rothenstein.

But such romantic musings were, for forty-five young men newly passed out of Sandhurst, far exceeded in excitement by the arrival of formal, typed letters from the Secretary of the Military Department at the India Office in Whitehall. 'Sir,' began the most momentous letter Walter Walker would ever receive, 'I am directed to inform you that it is understood from the War Office that your appointment to a commission on the Unattached List for the Indian Army in the rank of 2nd Lieutenant will be notified in the *London Gazette* on or about the 3rd February, 1933 ... The War Office will inform you of the British Regiment to which you will be attached for one year and you should hold yourself in readiness to embark for India on or about 8th February, 1933 ...'

This was soon followed by a long printed list of uniform clothing and equipment which a young officer should purchase from the military outfitters together with the reminder that 'Service dress or khaki drill and mess dress are to be worn on transports; swords and spurs are not worn on board ship or upon embarking.' Customarily, an allowance of £50 towards the cost of uniform was made and there was a free issue of a revolver, binoculars, a prismatic compass and such camp kit as '1 pillow, hospital pattern, hair, small', '1 bucket, water, General Service, canvas' and '1 chair, officers', camp'. But, in addition, the young officer was expected to be smartly fitted out with civilian clothes and a full list of the minimum amount of *mufti* was provided. This included such footnotes as, 'A mufti sun helmet can be bought cheaply in India. There are correct and incorrect forms, and the advice of a fellow-traveller who knows the East should be asked before purchasing' and 'Bowler or top hats are not worn in India.'

Finally, at the end of January, came orders to embark on the troopship *Somersetshire* at Southampton on 8th February, together with nine hundredweight of baggage so packed as not to exceed forty-five cubic feet, for passage to Karachi.

Prospective officers for Gurkha regiments usually spent their year's training with a British regiment such as the Rifle Brigade, or the King's Royal Rifle Corps, to which they were affiliated. But Edward Walker was now serving with the Punjab Regiment stationed at the garrison town of Multan and he had suggested that his younger brother try to spend his time with the 2nd Battalion, the Sherwood Foresters, which were also there. And so it was arranged that the journey beginning on the front-door steps of the house in Tiverton would end in the hot, dusty cantonment at a remote town in the Punjab.

Despite their uniforms and their Sandhurst training, the forty-five who set sail from Southampton were still schoolboys beneath their military bearing. When the tea bugle sounded in the troopship they scrambled into the saloon before the older subalterns and their wives to occupy the armchairs and stuff their mouths with biscuits.

The ship's adjutant had to warn them that unless they displayed more gentlemanly manners they would be punished by having to paint the ship while she passed through the Suez Canal, an ordeal to be sharpened by the heat of the sun and the mockery of the Egyptians on the shore.

Curbed by such occasional threats, they settled down for the six-week voyage, played deck games, fought in the boxing ring and took poorly-exposed snapshots of one another and of distant views of Gibraltar, Port Said, Aden and smoking tugs and trim, pale grey ships of the Royal Navy.

The voyage ended alongside a jetty at Karachi and the young men stepped ashore to begin their intense involvement with the East, which for some would be short and end on the battlefield or in a hospital bed and for others be unexpectedly long and rewarding. At this early moment their ignorance of India was immense. Walker probably knew more than most through his father's reminiscences, family letters and his fierce grandfather's autobiography. Before his interview with the Civil Service Commissioners at the age of seventeen, he had, on his father's advice, read a newly-published book about current Indian affairs: none of his questioners, who included a general and an admiral, had read it and had been quick to change the subject to tennis and show gratification when they heard that this boy admitted to less skill than his father. Even so, in retrospect, Walter Walker was to confess that he 'knew nothing' about India.

Indeed, few Englishmen did know much about the affairs of India. The officers who now commanded battalions and even companies at this time had spent their boyhood in the reflected dazzle of the British Raj at the height of its power and glory. At the turn of the century, when Lord Curzon had been Viceroy, it had seemed that the long peace which had followed the Mutiny would last for ever, excepting, of course, the inevitable frontier wars which offered splendid training for the young officers and welcome relief from social and sporting life. The great Durbars held in honour of King Edward VII and King George V, at which the hereditary Indian rulers had reaffirmed the homage they had accorded to Queen Victoria, seemed to confirm this comfortable belief on a grandiose scale. Almost certainly none of the forty-five new arrivals would have believed for a moment that the reason the apparently most-anglicised Indians spoke a curiously old-fashioned English slang was that, after the second Durbar in 1911, it was no longer fashionable among educated young Indians to try to pick up their European conquerors' mannerisms.

But already the groundswell of intellectual opinion – nationalist in India and sympathetically radical in the British Isles – was stirring the historical process and beginning to sweep it forward. By defeating Russia in 1905, Japan had proved that Europeans were not invincible and therefore not rulers by divine right. The distant but appalling view of Europe tearing itself apart in the First World War had given Indians an awareness of their rulers' frailty and of their own hopes for self-determination. British lip-service to eventual Indian rule of India was thus brought forward into the sharp focus of reality rather than possibility. Self-rule at last rose over the historical horizon.

The British declaration in 1926 that India was to be permitted to progress towards dominion status within the British Empire may have seemed a cosy liberal formula for satisfying Indian political aspirations while, at the same time, relieving the India Office of some responsibilities yet retaining the old political, economic and strategic advantages. But to the Indians it brought their hopes forward to a new frontier and their eyes could lift beyond it and ahead.

Walter Walker did remember from the time when his mother's pessimistic cousin had advised him to make his career in tobacco that he had been as pessimistic about the future of the Indian Army as of the Royal Navy. 'How long is the Empire going to last?' was the unthinkable question he had asked. And his uncle in the Indian Police, fresh from the disturbingly unexpected civil disorders, had doubted whether the Indian Army would still be commanded by British officers by the time Walter could expect to be commanding a battalion or a brigade.

But to the young officers in their tropical uniforms, Sam Browne belts and sun helmets, boarding the trains at Karachi for garrison towns, such thoughts were infinitely remote. They probably knew that the Indian National Congress, founded in 1885, had, since the Imperial Conference of 1926 become the vanguard of political revolt against any limitation of hopes for independence. They certainly knew that the leader of the independence movement – and Congress – was Mahatma Gandhi, for this extraordinary figure was the very antithesis of such rulers as Curzon. The slight, wizened and half-naked figure of Gandhi, with his symbolic spinning-wheel and his talk of non-violence but his undoubted control of immensely strong human forces, captured and alarmed the British imagination as it captured and thrilled the Indian.

The young officers knew that part of their time would be spent on new and unpleasant duties known as 'Internal Security'. Mass arrests of Congress leaders, demonstrations and riots had become part of the news from India since the end of the First World War and the massacre of demonstrators at Amritsar in 1919 by General Dyer's troops had brought racial mistrust to the most dangerous level since the Mutiny. But it was not to the teeming, uneasy towns that the newcomers looked with anticipation. Their youthful zest was ahead of them and their tin trunks of military paraphernalia, far away on distant frontiers facing understandable enemies, brave as they were cruel, opponents fit for the heirs of Clive.

Threats of civil disorder necessitated the allocation of one-third of the Army in India to internal security duties and a symptom of the general unease was that it was laid down that British should always slightly outnumber Indian soldiers engaged in such duty, usually at the ratio of eight to seven. But although the garrison of Multan was within reach of the cities of Karachi, Lahore and, indeed, Amritsar itself, it was essentially a strategic base in support of the garrisons of Quetta and Peshawar – to the west and north-west respectively – and these in turn faced the volatile North-West Frontier between the Himalayas and the Arabian Sea.

The train journey from Karachi to Multan lasted two days. Three others shared Walter Walker's boyish anticipation of their journey's end. Of these only one was to survive the war and retire, to become a controller of Civil Defence in England, many years after the other two had died, one on the North-West Frontier, the other with Major-General Orde Wingate, the Chindit leader, in his aircraft crashing into the hills of Assam. But now all was jollity and unclouded optimism.

In the restaurant car, the four young men struck up conversation with a friendly English middle-class couple. He, they were delighted to discover, was manager of Grindlay's Bank at Multan and so, they expected, would be a valuable point of social contact. During meals, this couple explained the protocol and etiquette required when dealing with Indians and recited some of the more simple and essential Hindustani phrases.

On the evening of the second day the train pulled into Multan and the experience of realities began. Baking on the plains of the Punjab, Multan had the reputation of being one of the most uncomfortably hot garrison towns in India. The temperature often exceeded 120 degrees Fahrenheit, rainfall from occasional thunderstorms rarely rose above seven inches in a year and, in the hot season, British troops left their cantonments for cooler hill stations.

Drill and tactical evolutions were customarily carried out during the mornings and the soldiers then confined to their barracks between mid-day and four o'clock in the afternoon to clean their weapons and equipment while Indian servants attempted to keep them cool by tugging the cords of slowly-swaying *punkah* fans and throwing buckets of water on to the thatched door screens. The young officers, who had Indian bearers to care for their uniforms and equipment, devoted these hours to the study of Hindustani.

The city of Multan had had a turbulent history, recorded in ruined ramparts and the tombs of its defenders. But its attractions seemed summed up in the Indian saying, 'Dust, heat, beggars and tombs are the four specialities of Multan.'

The welcome that first evening in the officers' mess of the Sherwood Foresters was exactly what any junior officer would expect from a regiment, other than his own, to which he was to be attached for training. At dinner, the officers in their scarlet mess kit either ignored or threw an occasional contemptuous glance at the new arrivals in crumpled khaki, only the adjutant showing interest and that in the fact that not only was Walter Walker's sun helmet a size too large but that the narrow leather piping around its rim had not been polished.

That night Walker did find an ally in his Indian bearer, who, knowing of the customary welcome, said comfortingly, 'Do not worry, sahib.' Then he took off his master's boots and unwound his puttees before squeezing half an inch of toothpaste on to his toothbrush and holding out his pyjama trousers at knee level. And, next morning, tea and hot shaving water were brought and the pressed uniform and polished boots and belt were ready and the bearer knelt before his master to wind his puttees before reverently handing him his hat and his regimental cane.

Any hopes that the bank manager, who had befriended him on the train, would provide a haven of relaxation for the new arrivals at his home were to be dashed. The four young officers had as-yet unopened packets of calling cards and had been instructed in the etiquette which insisted that they be delivered in threes, one for the husband, two for his wife. So, one evening after their Hindustani lesson, they dressed in blazers, grey flannels, suede shoes and sun helmets and set out by *tonga* cab for the bank manager's house. They were not invited in but deposited their cards and returned to the mess well pleased with their own knowledge of gentlemanly manners.

Next day the senior subaltern – a veteran of twenty-one years' service – asked the four to meet him in the mess anteroom. He there instructed them in the etiquette of calling; the number of cards to be left and the clothes to be worn. He then went through the list of those to be favoured with a call during an officer's first fortnight in a garrison town. The Political Service, the judge, the clergy, the doctors and the police were listed but there was no mention of the manager of Grind-lay's Bank. Walter Walker asked why this important figure had been omitted.

'But he's a box wallah!' exclaimed the subaltern. 'He's in trade! We don't call on bank managers!' Confessing that he had already called and left a card, Walker was warned that he must never invite this acquaintance to the mess. It seemed odd that the businessmen, traders and planters, whose activities the Army were there to protect and who might well have included members of their own families, should be so ostracised and it was to be embarrassing when the bank manager and his wife were met in the small social world of Multan. But, it seemed to Walker, there must be a practical purpose behind this baffling rule that would be revealed in due course and was not for him to question at this time.

The training and the discipline with the Sherwood Foresters was at least as demanding as that at Sandhurst but in India there was a sense of urgency, the constant possibility that an operational infantry battalion be sent on active service. Walker's new task master was his company commander, a Captain MacDonald Walker. A pugnacious officer, with a boxer's build and red hair and moustache and a quick temper, MacDonald Walker was much admired on the cricket field and in the squash court. Memories of school playing fields returned with the captain's roar of 'Catch it, you bloody fool!' or 'You bloody funk!' when a hard-driven cricket ball was dropped.

Two particular lessons were learned by Walter Walker from his namesake and both had to be learned with tribulation. One was night navigation. Captain Walker would hide a silver rupee some distance from the cantonment, then give each subaltern the compass bearings and numbers of paces that would lead him to it. They were then sent out into the night with orders not to return until the coin had been found. Hours spent stumbling over the rough, sun-dried mud around Multan implanted in Walter Walker the importance of not only overcoming the hindrance of darkness – or, for that matter, dense tropical jungle or Arctic blizzards – but of taking advantage of it.

The other was the tedious business of checking stores. The company commander chose Walker to supervise the checks of what were known as 'expendable' stores at the school for the regiment's children: pencils, pens, nibs, rubbers, pencil-sharpeners, rulers, exercise books and so on. The stores must be physically counted, ordered Captain Walker, with as much care as if they were rounds of ammunition or medical supplies.

At the school, the headmistress explained that, as was the custom, she had had all the listed items laid out in rows so that they could be counted easily. Indeed, if Mr Walker would like to make some random checks he could take her word for it that the numbers were exactly those that she had already listed for him.

After checking for an hour, Walker found no mistakes and so, accepting the mistress's word that her own counting had been accurate, he went through the lists more quickly and, an hour and a half later, was congratulating her on her accuracy and signing that the totals were as stated. At this moment, Captain MacDonald Walker walked into the schoolroom.

'Finished already?' he asked. 'It used to take me all day.'

Walker explained that he had found the mistress's counting so accurate that it had been unnecessary to count every single item himself. At this Captain Walker asked the woman to leave the room and turned on the subaltern in cold wrath.

'Would I be right in saying that my orders were crystal clear? Did I not order you to check every single item with as much care as if it were ammunition? You have disobeyed my orders. I require you personally to inspect and count everything. I want to know not only the number of pen nibs there are but how many of them have been used and how many are unserviceable. It *is* tedious. Life is tedious, battle is tedious. Report to me when you have carried out my orders.'

Walter Walker returned to counting the expendable stores and it took him all day. Two weeks later he was again counting but this time it was medical stores from the boxes designed to be carried into action by mules. Now every pill had to be counted and checked but the importance was more obvious. Constantly, stories circulated about ammunition stolen and the boxes filled with exactly the same weight in stones and there was, soon after this lesson had been taught, a story about fodder which had been dug from the centres of the great storage heaps in the mule lines and sold. It required no imagination to realise what would happen in battle if ammunition boxes were found filled with stones, or medical supplies ran out or the mules were starving.

Nine months after Walker had arrived at Multan, the Sherwood Foresters moved to Bombay where civil disorder was expected. This was not active service but – from the young officer's point of view – the next most exciting thing. After much checking and packing of weapons and equipment, the battalion made the move and, on arrival, at once set about studying the contingency plans that had already been drawn up and evolving their own. If violence was to break out it would probably begin with vast street demonstrations and escalate when these were contained by the police. So that, in such an emergency, the Foresters could rush to the points where the crowds could most easily be split by cordons, the officers had to become familiar with the city streets, both the wide boulevards and the markets where the crowds would gather and the connecting alleys along which they themselves would move to intercept them.

This would not be as exciting or challenging as skirmishing with the Pathans on the North-West Frontier but it was at least a taste of the realities of soldiering. There were plenty of reminders, too, that this was, for the Army in India, routine peacetime soldiering. The barracks occupied by the Foresters were, for example, across a street from the regimental tailors, the barber's shop, the ice contractor and the other commercial camp-followers. But soldiers were forbidden to cross the street, which was public, without decking themselves in walking-out dress: sun helmet, starched jacket buttoned to the throat, creased trousers, burnished boots and regimental cane.

The social life of Bombay also continued despite the threat of disturbance. The senior subalterns were much put out when Walker, who had grown into a strikingly handsome young man, was invited to the principal ball of the season by the pretty daughter of a senior civil servant. He was by the end of his first year in India the epitome of what the world imagined a dashing British officer should be. His slight frame had filled out with muscle, he acquired an even chestnut sunburn and his hair was well-cut and shone with a sparingly-applied and certainly-not-scented dressing. He had learned impeccable manners from his family and to this self-confidence had added a gay, relaxed but always attentive charm. His uniforms and mufti were, of course, immaculately cut.

The strict discipline of the Sherwood Foresters had not only been no impediment to Walker, he had even enjoyed mastering its little challenges and flexing his secondary military muscles by demonstrating just how smart and efficient a second lieutenant could be. Certainly his Confidential Report from his colonel could hardly have been more fulsome.

Describing him as having plenty of character and an attractive personality, the report went on to praise his exceptional zeal, initiative, tact, reliability and sound judgement. 'He is steady and has the courage of his convictions,' it concluded. 'If this young man goes on as he has begun, he has a grand future and may go far.'

There was no trouble in the streets during Walker's three months in Bombay and the time arrived for him to join the regiment which had accepted him when he was a seventeen-year-old schoolboy. The climactic meeting between the 1st Battalion, the 8th Gurkha Rifles and the young officer, who had been so carefully matured for their service, was to come at the end of long journeys for both. The battalion was temporarily at Loralai, a frontier station in Baluchistan, although it was shortly to move back westwards to the great military complex at Quetta. This was to be the rendezvous and, for Walker, the journey involved sea travel back to Karachi and another two-day train journey north but, instead of taking the right fork at Sukkur Junction for Multan, turning north-west towards the hills of Baluchistan.

After the heat and humidity of Bombay, Quetta was exhilarating. The sun could burn but the air had zest and there was a tang in the wind from the hills which could bring snow. The officers' bungalows in their groves of trees were designed to protect against heat and cold and there were reminders of home like open grates, chintz-covered sofas and armchairs and tweed suits hanging in the wardrobes.

Walter Walker was welcomed at the officers' mess of the 7th Gurkhas with more courtesy than he had been accorded at earlier arrivals among strangers. A small advance party from his own regiment had also arrived and they and he instinctively regarded themselves as belonging to the same family and the 7th Gurkhas as being agreeable cousins.

There was a week to wait before the 8th Gurkhas arrived, time to tour Quetta – the Indian town in the valley, the military cantonments on the slopes leading to the Staff College on a commanding bluff – but more important to see, for the first time, Gurkhas. 'The Little Men' as they were often admiringly known had been fighting alongside the British Army since the year of Waterloo and had long been as much of a legend in England as in India. It was said that there was a natural *rapport* between British and Gurkhas: mutual respect and liking between the soldiers and intense loyalty between British officers and the Gurkhas under their command. Indeed, commanding Gurkhas came so naturally to these officers that it could be argued that they were still employing the techniques of hearty leadership by example they had exercised at their public schools. The Gurkhas' simple virtues of unquestioning loyalty and courage seemed appropriately akin to the ideal qualities the British ruling class encouraged in their sons.

The first British contact with the Gurkhas had been, like so many with the 'martial races' of Asia, on the battlefield. Already, in the eighteenth century, the King of Gorkha, a mountain war-lord, had conquered the foothills of the Himalayas from Kashmir in the west to Sikkim and Bhutan in the east. Like most Asian hill-farmers, the Gurkhas, as they came to be called, regarded the raiding of the rich lowlands as their right. But the lowlands they raided were part of the expanding claim of the East India Company and, in 1814, their incursions were halted by the well-disciplined soldiers of 'John Company'.

The Gurkha's courage and toughness had so impressed the British that the Company at once raised four battalions of them for its own service. So began the long association. The stocky Gurkhas and their sharp, leaf-shaped kukri slashing-knives became familiar on the battlefields of a still-expanding Empire: the Burma wars, campaigns in the Naga hills and in Tibet and, of course, the interminable raids and counter-raids along the whole of the North-West Frontier from the Himalayas to the sea.

Even the slaughter in France and Mesopotamia during the First World War had not seemed to shake the Gurkhas' confidence in and love for the British for whom they died. The love was returned and was, perhaps, best expressed in a much-quoted passage from the introduction to *The Dictionary of the Nepali Language* by Professor Sir Ralph Turner, who had led Gurkhas in the First World War and had become the leading authority on their language.

'My thoughts return to you who were my comrades, the stubborn and indomitable peasants of Nepal,' he wrote. 'Once more I see you in your bivouacs or about your fires, on forced march or in trenches, now shivering with wet and cold, now scorched by a pitiless and burning sun. Uncomplaining you endure hunger and thirst and wounds; and at the last your unwavering lines disappear into the smoke and wrath of battle. Bravest of the brave, most generous of the generous, never had country more faithful friends than you.'

Love and gratitude apart, the Gurkha's rewards were few. His simple way of life and customs had been rationalised and adapted to Indian Army routine. Even his sparse home life and the raising of his family were regulated. During fifteen years' service a Gurkha rifleman could expect to have his family with him for only three years. He might return to Nepal for six months' leave, two of which would be spent in travel, every three years. Otherwise, a quarter of the Gurkhas in a battalion might have their families with them when not on active service but, as all the Gurkha officers, who commanded platoons, and most of the senior non-commissioned officers had their families with them by right, the rifleman spent most of his early adult years in celibacy.

The British officers who led them found their own lives ordered in much the same way. Their more comfortable way of life and customs had also been rationalised and adapted to Indian Army routine. An officer was not expected to marry until he was aged thirty and he, too, was granted home leave once every three years but for nine months, to include the long journey by train and P. & O. liner. The mess, which was the centre of his life, was like an Edwardian gentlemen's club, hedged about by many quaint traditions and customs and occasionally giving vent to its contained virility and zest with outbursts of boyish skylarking. When the officers took wives, they set up home under the shadow of the mess and their wives, while treated with a certain reverence, were always regarded as secondary in importance to the regiment and its 'Little Men'.

Finally Walter Walker's long wait was rewarded as the long columns of the 1/8th Gurkhas marched into Quetta and in the mess there were welcoming handshakes for the grandson of the Commanding Officer who had become part of their private mythology.

Any apprehension over the warmth of his welcome from the battalion was instantly dispelled. By any standards and particularly by Army standards, Walker was, socially, a most agreeable and, professionally, a highly promising young man. His Confidential Report for 1934 recorded that 'his attitude both towards his Superiors and Juniors is most pleasing, and he possesses a very pleasant Personality. His habits are strictly temperate.'

The mention of temperance was, as Walker was to find out, a reflection of a problem of this and, indeed, of most other officers' messes. There were bachelor majors in their thirties, who had fought in Europe and Mesopotamia as subalterns and were often said to have 'had a good war'. Some had returned to India and, instead of trying to create another such home as they had known as boys, settled for the pleasures of the mess and the bar. Unwritten regimental histories would be littered with sad tales of promising careers slowly ruined by alcohol and disappointed middle-aged officers fading away on their pensions and sometimes seeking a quick and violent end to their unhappiness.

While Walter Walker enjoyed a tipsy guest night as much as any other gregarious young officer, he heeded the warning of the sad superiors. But there was soon to be one occasion upon which all thoughts of such self-discipline would have to be abandoned. While his acceptance by his brother officers could have been taken as a foregone conclusion, a far more challenging and less predictable affair was his reception by the Gurkhas.

Meeting the Gurkhas, getting to know and understand them and finally winning and holding their loyalty and affection was a long process. The first sign of their reaction to a new officer was when he paid the first of his regular monthly social visits to the Gurkha Officers' Mess.

There were about twenty-five Gurkha officers with the battalion, a few more than the number of British, and although they held the Viceroy's Commission their customs had more in common with those they had always known than any copied from the British. Like the riflemen, they shaved their heads and, in the mess, ate their hot, spicy food and unleavened bread with their fingers. When Walker bicycled to their mess he knew that much would be expected of him. He was the grandson of a revered Commanding Officer and, on entering the mess anteroom, there was a portrait of Colonel Walker staring fiercely from the wall.

It was a lively evening with rum alternating with curry and it was clear that Walker met with his hosts' approval. It was said that he was the image of his grandfather and that he, too, would one day command the regiment. As the rum flowed, Walker's worries about the steadiness of his bicycle ride back to his quarters faded and the last recollection he had was the subadar-major and himself hurling darts at his grandfather's portrait. Expecting such an end to the evening, the Gurkha officers had arranged for the bicycle to be returned and its owner to be taken home to bed in a *tonga*.

The eight hundred Gurkha riflemen were to be met on the football field and the game became as much a part of Walker's routine as riding and tuition in Hindustani. Learning the language was a matter of urgency and not only because an examination had to be passed before going off on the first home leave after three years in India. Not until proficiency was attained in Hindustani or Urdu could an officer be taught Gurkhali.

Even before learning the language of his men, an officer was initiated into the mythology and traditions of the regiment. There was regimental history to read, the pictures and trophies in the mess to be recognised and protocol to be memorised. Little differences in dress and etiquette were prized in each battalion. The 1/8th, for example, wore a red hackle plume in their hats to commemorate their association with the Black Watch whose officers had, after one disastrous battle in the First World War, temporarily taken the place of their dead.

Although the battalion was preparing for an eventual return to active service on the Frontier, it had quickly settled down to an almost placid round of professional duties and social entertainment. Perhaps the symbol of Quetta at this time was the pair of starched shorts that was kept in every battalion and company office. An officer would dress each morning in another such pair but while bicycling to work these would slightly crease and had therefore to be replaced before their owner could be seen on duty.

Walker had naturally expected that his first encounter with the harsher realities of his profession – danger and death – would come either on the Frontier or in some riot-riven city. It was almost ironical that it should happen in Quetta itself.

As the seat of the Headquarters, Western Command and Baluch-istan District, the base of two Indian infantry brigades, the seat of the Staff College and the depot of various supporting arms and services – in all some twelve thousand Service men – Quetta seemed as safe as any stronghold in the Empire. Even the predatory tribesmen, who looked down upon it from the mountains, circled it warily, kept at their distance by patrols and light tanks.

The cantonment, where the three Indian, two British and two Gur-kha battalions lived, covered many square miles to the north-east of the town, spreading up the hillside on either side of a valley towards the brigade parade ground and the Staff College. On low ground, north-west of the town, lay the Royal Air Force station with its bar-racks, hangars and airfield. The town itself, built of brick, was compact, although its population had trebled during the past fifty years, with the growth of the garrison, to about seventy thousand. Together, the town, cantonment and air station covered twenty-five square miles.

The evening of 30th May, 1935, was mild and so regarded as the first day of summer. Many officers decided to sleep out of doors and Walker's bearer made up his bed beneath a mosquito net draped over crossed poles on the lawn of the bungalow he shared with another officer.

It was a pleasant evening for the guest night that Walker attended, returning, well-wined, to his quarters after midnight. Under the clear skies he slept deeply. Then at exactly three minutes past three in the morning he woke with a jolt to a nightmare of an express train roaring and screaming down upon him. The bed rocked, the mosquito net col-lapsed upon him and he opened his eyes to see the corrugated-iron roof of the bungalow heaving violently. Struggling from the folds of netting, he ran to a cherry tree across the trembling lawn and heaved himself into its branches. The ground could open in earthquakes and he and his companion had instinctively taken to the trees

The roaring and rumbling rolled down the valley towards the town and slowly died away. Then the wavering light of fires began to light the sky to the south-west above the town and across the cantonment came sharp, urgent bugle calls sounding the alarm. Walker pulled on some clothes, found his bicycle in the darkness and rode off towards the Gurkha lines.

Quetta was prone to earthquakes, but not even two severe shocks four years earlier had prepared the city for the disaster of that night in May. So shattering was its effect that even the official account published by the Government of India three months later seemed infected with near-hysteria.

'In Quetta the scene is indescribable,' the narrative ran. 'A whole sleeping city; men, women and children crowded in small rooms, large rooms and corridors, on roofs and on pavements; animals tethered in tiny enclosures; a mass of humanity sleeping in a house of cards, each card of which weighed a ton....

'The street lamps are alight; police constables patrol their beats ... then, a sudden dynamic convulsion; a surging implacable wave roared over the surface of the earth and made Quetta – in less than half a minute – a shambles, a catacomb.... The wave passed through Quetta leaving few fissures, no boiling mud or other outward marks – except a heap of debris and ruins, the city which was once Quetta.

'For the next ten minutes no one knew what had happened. Then humanity began to re-assert itself and our imagination will help us to visualise the heart-breaking scenes....'

There was no moon and the electricity supply had failed, so Walker and his Gurkhas and the other troops in the cantonment cautiously made their way downhill towards the light of fires where broken oil lamps had ignited splintered timbers. As the first light of day came up over the ridge of mountains, they saw a vast cloud of dust hanging over the ground where Quetta had stood.

The town itself had been destroyed, its houses shaken down into heaps of rubble. It would never be known how many people were killed in the twenty-five seconds of earthquake, or how many died later, buried under their homes, but the Government finally put the total at between fifteen and twenty thousand.

The RAF station had also lain in the path of the worst shock wave and had been destroyed, the barracks reminding one officer of 'a village in the front line in France after three years of war'. Of a total strength of six hundred and fifty-six, British and Indian, on the air station one hundred and forty-eight were killed and about two hundred injured.

Walker's company made straight for the Residency where they had mounted the Agent-General's guard. Sir Norman Cater himself had escaped but almost his entire staff and guard had been killed beneath the collapsing house. Some of the Gurkhas had been trapped under the masonry by their rifles to which they were chained as an added precaution against stealthy rifle thieves.

By six o'clock Cater and Major-General Henry Karslake, the General Officer Commanding, had set up an emergency headquarters on the lawn of the ruined Quetta Club and troops were allotted areas for rescue work. The 8th Gurkhas were sent to the south-east of the town and there they began to dig. But only one in ten of those found in the rubble was alive.

The inevitable consequences of such disasters followed. The corpses began to smell and rescue parties were issued with face masks to enable them to work without continually vomiting. The Indian dead were carried to vast funeral pyres and the living were moved away from the ruins which were surrounded by barbed wire and sentries to contain potential disease and exclude looters. Refugee camps were set up on the racecourse and polo ground and Walker and his Gurkhas were sent to hold back huge crowds besieging the railway station in the hope of escaping by train. Baluchi tribesmen were circling closer to the town in the hope of loot and looters who were caught were sentenced under Martial Law to flogging.

Finally, relief supplies began to reach Quetta and the worst was over. For many young officers, including Walker, it had been their first experience of an emergency and the Army made its point that this sort of affair was all part of a soldier's expectations by taking troops off relief and guard duties as soon as possible to prepare for a full-scale ceremonial parade before the Viceroy of India.

As a precaution against further earthquakes, living quarters in the cantonment were rebuilt with walls of mud brick and roofed with canvas. Then, to ease pressure on accommodation before the cold weather began, the garrison was reduced, both battalions of the 8th Gurkhas being sent away, the 2nd to Loralai on the Frontier and the 1st to the regimental depot at Shillong on the far side of India.

The administrative capital of Assam, Shillong was the hill station for Calcutta. When the hottest weather began in the Ganges delta as many of the British who could work elsewhere, or could afford the time and expense, travelled three hundred miles to the north-east into the pine-forested hills and tea gardens of Assam. Shillong itself lay at nearly five thousand feet above sea level, the temperature rarely rose above eighty degrees Fahrenheit and the average rainfall was among the highest in the world.

For Walter Walker, Shillong and Assam were familiar as part of his family history. Their tea garden 'Tara' was here and it was from Shillong that one of his uncles had commanded as Commissioner of Police. The town with its churches, club, English shops and hotels and the wooded hills around it were more reminiscent of Scotland than of the harsh India he had seen up to this time. There was a busy British social life at Shillong with dances at the club each week and polo as the most popular game.

There was now more than old family connections to remind Walker of home. In 1936, three years after he had disembarked at Karachi, there was home leave to be expected. Too impatient to spend the first six weeks of his leave on board ship, Walker decided to delve deeply into the £500 travel allowance, which was all that was due to him for his entire Army career, on flying home and returning cheaply by sea. Imperial Airways operated a flying-boat service from Calcutta but this was a sufficiently remarkable means of travel that, by using it, Walker was assured of an admiring interview with the local newspaper in Devon.

Neither 'Tara' nor its inhabitants seemed to have changed in the three years; his father genial and proud of his son, his mother worrying but proud. But Walter had changed. The eager, boyish young man who had left, returned, his sun-burned face marked with the first lines of experience and a dashing military moustache grown on orders from the adjutant. Soon he asserted his new status by buying a small Morris car and driving off to tennis parties and, later, to hunt with the Tiverton foxhounds.

The one reminder of his future prospects was a fortnight's drill course at the Rifle Brigade's barracks at Winchester. The news from Germany, Italy and Spain indicated a troubled future for Europe, but India was a long way away and had problems of its own, although none, it seemed, that could not be contained, or solved, by the King-Emperor's armies.

Then the nine months' leave was nearly over and, little of his travel allowance remaining, Walker planned to return as cheaply as possible by sea. There were sad farewells at 'Tara' because there could not be another reunion for three years. That would be in 1940.

Shillong was reached in time for Walker to join the battalion in their camp out in the hills for concentrated training in tactics and weaponry. Again, in due season, came the return to the depot, polo twice a week and, afterwards, a dance at the Shillong Club.

For the unmarried officers of the 8th Gurkhas, which meant virtually all the officers under the age of thirty, attendance at club dances was a pleasantly masculine, self-satisfying routine. At such hill stations unmarried British girls could meet eligible bachelors with a suitable military or civil service background. So they gathered at Shillong, not only from Calcutta in the hot weather, but also from England. They were known – as were their Navy-minded contemporaries in Malta – as 'The Fishing Fleet'.

On these evenings, the single officers of the 8th Gurkhas would arrive at the club dance wearing gardenias in the buttonholes of their dinner jackets. After a glass of whisky and soda in the bar, they would stroll into the ballroom – usually just after 'The Paul Jones' dance had demonstrated agonisingly that the women outnumbered the men – and cast coolly appraising eyes over the unattached girls, 'The Wallflowers'. Then, without making any approach to them, the young men would stroll back to the bar for a second drink before condescending to join the ladies.

Among the girls at Shillong during the 1937 season was one who did not belong to 'The Fishing Fleet'. Beryl Johnston's father was a teaplanter at Pabhoi in Assam and, to her, Shillong was as much a centre for sociability and shopping as Cheltenham had been before she came out to India. Moreover she was an outstandingly beautiful girl – tall, graceful, with fine eyes and a disarming smile – and had neither the need, nor the inclination, to curry favour with the smug young officers at the club. Indeed, at one dance, she saw the blades of the 8th Gurkhas stroll into the room for their first inspection, considered them 'too pleased with themselves by half' and noticed that the most handsome, who appeared last with studied carelessness, was a Lieutenant Walker.

Later, a dance in fancy dress was held at the club and Walter Walker invited Beryl Johnston to join him. She had thought him 'pretty casual' but now he showed himself to have a pleasant manner and she accepted. They danced together for the rest of the evening. At the end of the dance they slipped away and drove her father's big American car on to the golf course. There they remained so long that when they tried to drive away the wheels skidded and the chauffeur was to report to Beryl's father that the car must have been parked somewhere it should never have been. But her father need have had no worries about Lieutenant Walker's intentions. Beryl Johnston and Walter Walker were engaged to be married.

But Walker was only aged twenty-four and it was taken for granted that they must wait a minimum of three years before marriage.

Lieutenant-Colonel Bruce Scott, a strict disciplinarian and 'a great Frontier man' fresh from the Tochi Scouts, had taken command of the 8th Gurkhas and had chosen Walker to be his adjutant. He made it clear that this job would be so demanding that marriage would be out of the question. Then Colonel Scott went home on leave. His temporary replacement, who was earmarked to succeed him, had himself married young and announced that he had no objection to the marriage taking place as soon as desired.

Walter had written to his mother about his engagement, expecting an enthusiastic reply. But his grandmother and Beryl's had known each other at Cheltenham and they were not friends. So not only did Mrs Walker disapprove of his choice – although she had not met the girl – but she told her son that she had written to Colonel Scott at his home in England and he had replied that of course there was no question of Walter being allowed to marry so young.

Before matters could be taken further orders reached Shillong that the battalion return to Quetta to prepare themselves for service on the Frontier. In March, 1938, the move was made and, soon after, Colonel Scott returned from leave. Walker was sent for and told that his marriage would normally not be approved. As adjutant, his work must take up virtually all his time and certainly he could not allow domestic life to occupy much of his attention. Moreover, the battalion was likely to be on active service for three years and the adjutant's turn for what little leave there would be always came last. Even so, said Colonel Scott, the marriage had been approved by his replacement and therefore, with reservations, he would confirm this approval on condition that Walker realised that his future would depend upon his success as adjutant.

On 9th November, 1938, Walter Walker and Beryl Johnston were married at St John's Church, Calcutta. None of his brother officers could accompany him across India for the occasion and his best man was a friend who was commanding the Governor of Bengal's Bodyguard. There were the usual minor disasters. The officiating clergy insisted that the fee for the choir be paid in advance and, as the best man had forgotten to bring any money, this had to be borrowed from the grooms in the Governor's stables. And the archdeacon remarked that he did not care for the flowers decorating the church because 'they always remind me of funerals'.

After the worry and the scramble across India, the honeymoon at Puri on the east coast of India was idyllic. Surf washed sandy beaches and there were beach huts roofed with palm leaves. This could last only a fortnight before setting out for Quetta and visiting Edward Walker and his wife at Lucknow on the way. When they arrived in December, Quetta was under snow and the Walkers' first home was a bungalow near the Staff College that had not been badly damaged by the earthquake. Yet Walter Walker was not often there. Colonel Scott, seeing the lights on in his adjutant's office late in the evening, would enter and tell him to go home to his wife.

The friendship between Scott and Walker became closer as a result because the latter, deeply distressed to have gone against his colonel's wishes on early marriage, was determined to prove that this would not impair his efficiency and would give him an added sense of duty and dedication. The two men and their wives established a happy relationship that was to flourish for more than thirty years.

The Walkers had had little more than three months of married life when the battalion's time came to move up to Razmak, the main operational base on the North-West Frontier facing Waziristan. In March, 1939, the 8th Gurkhas were packed, ready to move and Beryl Walker prepared to cross India again to stay with her father in Assam. This was a pattern which Beryl – like her contemporaries and predecessors – would have to accept. There would be no settled home until her husband retired – or died.

During the next two or three years, the couple could expect to see each other for a few weeks at the most. Even that was now doubtful. Since Walker had been in England three years earlier, much had happened to change the future outlook. In Europe, the Munich crisis of the previous year had showed that there were no obstacles to further expansion by Germany, just as Italy had been allowed to seize Ethiopia and both Fascist dictatorships were able to help the Nationalists win the civil war in Spain. With Italy now a major colonial power in North and East Africa, any conflict between the dictators and the European democracies would almost certainly spread to the Middle East and involve the Indian Army. But, for the moment, the prospect of service on the North-West Frontier, awaited so long, was enough to occupy the mind of any young officer in the employ of the Government of India.

At this time the chancelleries of the world were much impressed by an item of news from England. In debate at Oxford University, the Union Society had declared by vote that they – presumably the sons of the ruling class, the flower of England – were not prepared to fight and die for King and country. This, it was widely believed, demonstrated conclusively that the British were decadent and enfeebled and that, if war came, were no longer the men who had conquered their empire.

A more significant symptom would have been seen at Quetta and that was only one of many such. A young couple, newly married and much in love, freely deciding that they must part for years so that the husband could carry out the duty to which he had committed himself. Even more impressive would have been the knowledge that, as the troop train pulled out of Quetta, that young officer would not have considered for a moment that his action was anything out of the ordinary.

2

Frontiers of Empire

The journey from Quetta to Razmak was a transition from theory into practice. The former had been dominated by thoughts of the Frontier: its politics and strategy in the Staff College; its tactics in the lecture huts and on exercises; its presence in the view of distant mountains. But the latter dominated, or tried to dominate, the Frontier of which it was a part.

Razmak was deep in hostile territory. The troop trains stopped at Bannu, a garrison station with its own brigade on active service. No women were allowed farther forward than Bannu where a few of them were allowed to stay in the unfortunately-named Abandoned Wives Hostel. From Bannu to Razmak, troops moved by motorised road convoys and it was at once apparent that any movement in this direction had to be planned and executed as an operation of war. Live ammunition was issued, armoured cars trailed plumes of dust among the packed lorries and, all at once, the invisible presence of enemies made itself felt. Soldiers in the crawling column looked up at the bare brown hills and could sometimes see the flicker of a heliograph or a tiny movement as piquets moved up a spur to command the heights.

On the Razmak road it required little imagination to feel the whole weight of the Frontier pressing down upon India. Like a shallow Norman arch, it spanned three thousand five hundred miles from the Arabian Sea to the Bay of Bengal. In the east, the first six hundred miles of frontier were with Burma and seemed impregnable, for surely no modern army could advance through those jungles, forests and man-grove swamps. In Sikkim there was an entrance to India – the trade route from Lhasa – but there could be no conceivable threat from Tibet or China. But then, for one thousand five hundred miles, stretched the most awe-inspiring and impregnable barrier of mountains in the world: the Himalayas. This range, over one hundred miles in width, its peaks standing twenty thousand feet and more, defended India with more invincibility than any army. Here was the highest mountain in the world, Everest, which had not yet been climbed, although two Englishmen might have done so, but they had not returned from the final assault; and now Englishmen had been the first to fly over the mountain. When the British made their attempts on Everest – and nobody doubted that the Union Jack would be the first flag to fly from its summit – they were accompanied by Sherpa porters from Nepal. These were of the same stock as the Gurkhas who accompanied their British officers to the Frontier in a combination of loyalty, courage and skill that seemed unshakeable.

West again lay Kashmir where two ancient trade routes lay through a wild jumble of high mountains into Sinkiang but these, too, were thought to present an unacceptable obstacle to any invader. It was where the Frontier ran south-west from the mountains of the Hindu Kush towards the deserts of Baluchistan that stretched to the Arabian Sea, that the danger lay. Where Afghanistan presses against India was the country known as the North-West Frontier.

Through Afghanistan ran the invasion routes into India. From the north in descending order came the great gateways of India: the Khyber, the Kurram, the Tochi, the Gomal and the Khojak and Bolan passes.

Astride the frontier and the passes, living in the mountains, were the Pathans, the warlike tribes. Some three million of them, living in a long strip of wild country of some twenty-five thousand square miles, untamed by any outside administration, ruling themselves with fierce independence. Of their half-million fighting men, nearly half were armed with rifles and it was their achievement that when the most powerful empire yet known to the world wished its servants to travel this road from Bannu to Razmak it could only do so by using the power of its immense military machine.

Were it only these formidable Pathans that need concern Quetta, New Delhi and Whitehall, peace-keeping operations on the North-West Frontier would, like other colonial wars, have been regarded as providing 'splendid training for young officers' and little more. Yet beyond the Pathans' homeland and beyond Afghanistan itself lay Russia.

The only challenge to British domination of India, once the French had been defeated in the eighteenth century, could come from Imperial, then Soviet, Russia. The tribes of Asiatic Russia had swept down the invasion routes into India before and it seemed possible that they might come again, whereas no European invader had marched overland into India since Alexander the Great. So secretive were the Russians, so closed their society, so inscrutable their sombre, sometimes barbaric, view of the outside world, that their intentions were difficult, if not impossible, to assess. Ignorance bred fear and pessimism and both fed on the undoubted fact that Russia, like Britain, desired India.

Thus on the North-West Frontier was played what came to be known as 'The Great Game'. Spies and explorers, political agents and soldiers set out into the wilderness, probing for the invader they expected to find poised for aggression. When there was no invader to be found, then the tribes could be suspected of being in the invaders' pay and of preparing his way. Thus, an ambush on the road from Bannu to Razmak, or the rising of a few hundred Pathans near one of the gateways to India always set intelligence officers wondering about the Russians. When, the saying went, will Ivan come?

When the 8th Gurkhas moved up to Razmak, the North-West Frontier had been in a virtual state of war for a hundred years. Between major campaigns – beginning with the First Afghan War of 1839 – had been the murders and the punitive expeditions, the ambushes and the skirmishes.

Columns of British and Indian troops would regularly plunge into the uplands to burn a village in retaliation, or 'open' a road dominated by snipers. Once the columns had been marching infantry escorted by lancers and supported by guns slung on the backs of elephants, camels and mules. Now there were motorised infantry, armoured cars and bombing aircraft.

The fortress of Razmak was itself the result of increasingly bold British strategic thinking which had resulted in 'The Forward Policy'. This meant that, instead of launching small invasions of Pathan territory, the Army in India would itself move forward and operate from permanent bases well inside that territory. In 1922 it was decided to establish a large fortified base deep in Waziristan, the country of the Wazirs and Mahsuds, the most troublesome of the tribes, and this was at Razmak. Here were to be stationed a brigade of six infantry battalions, with a brigade at Bannu and another at Wana in support.

In all, there were nine brigades operating on the North-West Frontier supported in turn by the Field Army of four divisions and four cavalry brigades held farther back in India. Seven of the eight Royal Air Force squadrons in India were also stationed along the Frontier.

When Walker reached Razmak with his battalion in March, 1939, the Frontier had been in a state of turmoil for three years. Never had the tribes seemed more united or aggressive. This had been brought about by one man and might have been ended without further bloodshed had one of the many shells and bombs intended for him found its mark. This was the magnetic character, who had become almost as much a subject of a new mountain mythology in the British Isles as on the Frontier, known as the Fakir of Ipi.

An unknown holy man named Mirza Ali Khan had been transformed into the Champion of Islam by one of those apparently minor and inconsequential incidents that have so often triggered widespread trouble. In this case it was the abduction, in 1936, of a Hindu girl by a Moslem youth and the subsequent legal tangles. The religious issue rapidly became a political one and the Fakir of Ipi emerged, not simply as the leader of a single tribe, but of all Pathans and his enemy, not simply the local British Resident, but the Government of India itself. And, in the years that followed he was to prove himself a bold and elusive leader, worthy of the fighting men he led.

Grandfather: Colonel Thomas Walker (1837–1902), Commanding Officer, 44th Bengal Native Infantry (8th Gurkhas).

Grandson: Lieutenant Walter Walker of 8th Gurkhas in 1936

Walter Walker (*standing, second from left*) with his brothers and sisters at 'Tara' before leaving for India

Gurkha machine-gunners of the Razmak Brigade in action on the North-West Frontier in 1939

Captain Walker in sleeping quarters while 'on column'

Razmak, from which the main operations against the Fakir were mounted, was both fortress and encampment, laid out in a square, like a Roman camp, its stone wall perimeter fortified with barbed wire, blockhouses and massive, flat-topped towers. Within were solid, stone-built, single-storey barracks, messes, headquarters blocks and a bazaar. Without, bare mountains rose steeply from the valley floor and, had the Pathans possessed artillery, this would have rendered Razmak untenable.

The two principal tasks of Razmak Brigade were to keep open the road to Bannu and to set out on punitive expeditions. The first was a regular, routine task, which, in 1939, was carried out twice a week. The second, launched when necessary, began much as a road-opening operation, but the initial tactical deployment led to an attack, or a cordon and search, as its climax. Operations might last anything from a few days to a week and nobody involved was tested more than the adjutants. As adjutant of the 8th Gurkhas, Walker had never been busier; when the battalion was not out on an operation it was actively preparing for the next one.

By wartime standards, the 8th Gurkhas went into action impeccably turned out: shorts starched and pressed, wide-brimmed hats set at the same jaunty angle. But once operations began, the work was arduous and demanding, both physically and mentally. The essence of Frontier warfare was to command the high ground, which could mean first having to dislodge the enemy from well-sited, strongly-held positions. Although the Pathans were generally armed only with rifles – often copies of the ·303 Lee-Enfield made in their village workshops – they were masters of the skills of 'fire and movement' and crack shots.

As the columns moved forward with their infantry, pack mules and armoured cars, piquets were sent to the tops of the commanding heights, making their way to the crests up the length of spurs, so as to be exposed to fire from above for the minimum possible length of time. Once the high ground was occupied, the piquets signalled to the column below by heliograph, or semaphore, that it was clear of enemy and the road immediately ahead was safe. Other piquets were now sent up to the next heights and, as the column passed, the first parties were withdrawn. This withdrawal was the most dangerous movement for the Pathans knew where the piquets had been and would try to reoccupy their positions and shoot at them as they scrambled down the mountain.

This was a time of tensions. Beside the road, officers of the main force would watch the crests through binoculars and mountain artillery and Vickers machine-guns would stand ready to open supporting fire. When the signal 'Prepare to retire' had been made and all was ready, the single word 'Retire' would be spoken by the colonel to the adjutant and by the adjutant to the signals officer. When the piquets received the order, they would sprint down the mountainside. If the enemy was to attack it would be now and, at the first rifle shot, machine-guns would rake the heights and a platoon, sent forward for the purpose, would at once put in a counter-attack.

In the summer of 1939, Walker first heard the snap of bullets and the whine of ricochets in the mountains of Waziristan. The first time his behaviour in action was to be noted was on a punitive expedition against a Pathan village called Mir Khan Khel Kalai in August. Typed on a buff Army Form W-3121 was the brief commendation: 'It was greatly due to the energetic staff work of Lieut. W. C. Walker that the 1/8 G.R. were successful in reaching very difficult objectives during the night. At light next morning, under heavy sniping fire, he helped to supervise the changes in the dispositions of forward troops.'

And, in the same area, a few weeks later: 'Lieut. Walker's energetic work as Adjutant greatly facilitated the task of the Rear Guard Commander. On several occasions when the Rear Party was difficult to withdraw, he went forward under heavy fire and personally conducted the withdrawal of platoons in difficulty.'

At the end of the year, the battalion retired to Razmak for Christmas. There were masculine festivities: boxing for the riflemen and a horse show for the officers, in which Walker and his charger won first prize. On Christmas Day, the British Officers put on their best tweed suits and suede shoes to become hosts to the Gurkha officers, all dressed in pillbox hats, regimental ties, rifle-green blazers and carefully-pressed grey flannel trousers. While the band played they sat outside in the wintry sunshine laughing and becoming pleasantly merry. Singing carols at Razmak at Christmas in 1939 was to be intensely remote from the world. In September, Britain and France had declared war on Germany and, although serious fighting had not begun in Europe, the Frontier now seemed forgotten by all but the soldiers in the camps and columns and even their conversation now turned more readily to von Rundstedt, or Lord Gort, or General Gamelin, than to the Fakir of Ipi.

In the autumn, Colonel Scott had left Razmak to take command of a new brigade being formed in Upper Burma – a quiet part of the Empire with little or no prospect of active service, it seemed – and had been succeeded by Lieutenant-Colonel Eric Langlands. For a few months Walker became a company commander with direct tactical responsibilities in the field. A testing time was soon to come.

During 1939, the battalion had fought several brisk skirmishes with the Pathans, two of them in covering the withdrawal of the Leicestershire Regiment from untenable positions. In March, 1940, another somewhat similar action was fought over the same ground as the last of the previous year's actions, but this time in support of the Suffolk Regiment.

It was typical of many small battles fought to cover the withdrawal of a high piquet. This was coming down from a hilltop to the left of a ridge, known to be held by Pathans, and two of Walker's platoons, which were covering the move, were also being withdrawn. Then four shots were fired from near the position the piquet had evacuated and a Gurkha was wounded. Walker at once threw the deployment into reverse to recapture the high ground. This was done, but another rifleman was wounded. Walker now called for artillery and machine-gun fire not so much to cover any further advance, or a new withdrawal, but the rescue of a badly wounded Gurkha, who was lying in the open between his and the Pathans' positions, and three volunteers were crawling forward to reach him. Walker, directing operations from a sangar – a low breastwork of loose stones – came under accurate fire, one shot hitting the light machine-gun beside him, splinters cutting his hand. As firing became general, the rescue party reached the wounded man. But in retrieving him, a Gurkha machine-gunner was killed and two more wounded. The general withdrawal on Razmak could now be continued.

Many such actions were being fought on the North-West Frontier at this time but what was to be important for Walker's future prospects was that his efficient controlling of the battle had been watched by both his colonel and his brigadier. That night Colonel Langlands wrote to Beryl Walker: 'Just a wee note to tell you that Walter has really distinguished himself today. He was in command of our rear party during a withdrawal on Razmak when one of our piquets was heavily fired on…. He really was excellent – organised things fast just as if he was on peace training in Shillong. The Brig. was with me – watching him and says he has never seen a show – under heavy fire – so well directed. Tomorrow I shall do my very best to get something for Walter. He really is a first-class soldier – you must be so terribly proud of him.'

About three weeks later, she received another letter, reading: 'By now you will have heard from Walter about his show for which he was put in for an MC. I thought I should like to let you know how sorry I and everyone else in the Battalion was that the recommendation was stopped by District. If anyone ever deserved the medal Walter did and it's most shockingly bad luck that the District Commander should have seen fit to do as he did.' Instead of the Military Cross, Walker's name appeared in a long list of those 'brought to notice by His Excellency the Commander-in-Chief'.

A few weeks after this skirmish, the German armies invaded France and the Low Countries and, more significantly for the Indian

Army, Italy was at war with Britain. With Italian colonies stretching along the eastern shores of North Africa and from the Sudan across Ethiopia to the Indian Ocean, it now seemed likely that the Indian Army would be called upon to carry out the grand strategic role for which it had been for so long prepared. Along the whole length of the great 'Lifeline of Empire' – the sea route from Britain, past Gibraltar, Malta and Cyprus, through the Suez Canal, then via Aden to India and beyond – the most vulnerable pressure-point was the Canal and the Red Sea and this might be the next theatre of war.

Walker's zeal had commanded so much attention that he was now appointed staff captain to the Razmak Brigade with an enormous range of responsibilities and serving directly under the Brigade Major. This was no desk job, for one of his tasks was the organisation of the temporary camps occupied and evacuated when the brigade was on operations. But to his despair his first duty as staff captain was to decipher a signal ordering the 8th Gurkhas to prepare to leave Razmak instantly for an unspecified destination.

As the long column of heavily-laden lorries rolled down the road to Bannu and the railhead, there was much talk of the Middle East, or East Africa, as their destination. In fact, it was to be Dehra Dun, the military base one hundred and fifty miles to the north-east of Delhi below the Himalayas. For Walker this was a bitter moment. For years he had trained to go to war with the 8th Gurkhas. He had been blooded on active service with them but now – so it seemed – they really were going to war and he could not go with them. As the column disappeared down the road, he stood at the gates of Razmak, with tears in his eyes.

There was now a new tension in the air. The wireless in the mess, or in the signals truck, was always tuned to the news broadcasts and in the remoteness of Razmak and the mountains of Waziristan, Captain Walker heard of the German occupation of Europe, the air fighting over Britain and the long agony in the Atlantic. It was frustrating to have to put all effort into fighting a few thousand ragged tribesmen instead of standing up to Nazi or Fascist armies, but, so intense was activity on the Frontier, that there was little time to brood.

By this time, Razmak Brigade had been built up into what was virtually a division and Walker found himself responsible for an immense variety of duties. With only the help of one Indian clerk, he had to handle the supply of rations and ammunition, for example; all legal problems, building projects and the relief of units. When 'on column', it was he who drew up the order of march, laid out the temporary camps (usually in twenty minutes) and supervised their efficient and tidy evacuation.

The siting and laying-out of these camps had to be carried out with that punctilious attention to detail in which Walker excelled. Whether the site had been used regularly by columns, or whether it was new, the procedure was the same. The column would stop at two o'clock in the afternoon to begin the long routine. First, the chosen ground was occupied and searched for mines and booby-traps. Piquets were sent out to any commanding heights. Then, to the plan sketched by the staff captain, slit trenches would be dug and sangars built. When all was ready, there would be a practice stand-to and, finally, all but the piquets would be withdrawn within the perimeter before dusk.

The slit trenches provided shelter from sniping and this would begin as darkness came on, the bullets often snapping over the sleepless soldiers until dawn. Dry stone walls had been built to protect the mules but the beasts were often hit, bullets sometimes passing clean through the body without touching the vitals so that all that would be visible in the morning would be a small pool of blood on the ground and a tiny puncture on each side of the mule.

While striking camp next morning, the staff captain had to make certain that all rubbish had been buried and the latrine trenches filled in, because the site might soon be used again. The column could then move on, and the process repeated the same afternoon.

If Walker needed another lesson in the dangers of the slightest disregard of instructions he was to be given a fearsome reminder.

A column from Razmak was moving through the mountains and the Pathans were known to be hovering beyond the crests in force. Particularly strong piquets were being sent up to cover the progress but one of them did not withdraw down the mountainside with the speed and efficiency this inherently dangerous manoeuvre demanded.

Twenty or more Indian sepoys had made up the piquet and, seeing and hearing nothing of the enemy, they walked rather than ran down the mountain and, instead of taking the more difficult route along the sharp edge of a spur, chose the easier way down a dry ravine. It was there that they were ambushed by the Pathans.

Forty-five minutes later, Walker saw an exhausted sepoy stagger towards the column. He was, he said, the only survivor of a massacre. At once a counter-attack was mounted and Walker went forward to retrieve the dead. The enemy had, of course, disappeared but what they left behind made that day in the mountains unforgettable. The sepoys all lay dead in a welter of blood. The Pathans had gone to work with their knives and the bodies lay, strewn on the rocks, beheaded, castrated, their eyes gouged out. It was, of course, known that this was a Pathan custom but it was rare to see one mutilated corpse – so important was the recovery of the wounded – but to see some twenty was an appallingly effective lesson in the dangers of relaxing discipline and tactical technique for an instant.

At a time when twenty soldiers might be killed in Europe by a single shell, such a loss might have seemed trivial. But the fact that the Indian Army could suffer such a defeat at the hands of mountain guerrillas threw a shock wave from Razmak, through Quetta to New Delhi itself. Later, when Walker heard himself described as a disciplinarian and a martinet, his mind's eye would see again that terrible blood-stained mountainside.

Now Beryl was able to come within a hundred miles of Razmak and, for several months, she stayed at the Abandoned Wives' Hostel at Bannu. But her husband was only able to visit her there twice, once when he was ill with jaundice.

News of the war in Europe and the Middle East dominated all thoughts of the future and the officers at Razmak assumed that, sooner or later, they would find themselves fighting over their fathers' old battlefields. Certainly few imagined that, before the end of 1941, they would be strategically stabbed in the back, facing an enemy far more formidable than any they imagined. The Japanese attack on Western possessions in the Pacific and in South-East Asia came as a stunning surprise.

The possibility of war with Japan had, of course, long been studied by contingency planners in Whitehall and in Singapore. It was seen as a threat to Malaya, the East Indies and, perhaps, strategically to Australasia but certainly not to India. To reach India, the Japanese would have to advance through Burma, crossing the grain of the country over high mountain ranges and across wide rivers, through dense jungle and probably torrential rain, without the benefit of anything worthy to be called a road system. Therefore India could safely look to the north-west as it always had, beyond the Pathan hills to the Middle East and beyond.

As late as August, 1940, the Chiefs of Staff Committee sat in London to review the threat in the Far East and concluded that while the Japanese might occupy Siam, this would only lead to the possibility of an air threat to Burma and that an invasion of the country remained a remote possibility. Malaya, it was considered, was so great a prize for the Japanese that all their offensive effort and all the British Empire's defensive effort must inevitably be committed there. The defence of Burma was therefore to be the responsibility of the Commander-in-Chief, Far East, in Singapore.

Then, on 8th December, 1941, Japan launched her war of conquest with the surprise attack upon the American naval base at Pearl Harbour. In quick succession came the assaults on the American islands, the Philippines, the Netherlands' East Indies and on Hong Kong and Malaya. But even when strong Japanese forces were known to be moving across Siam towards the Burmese border and after Rangoon suffered its first air raid, towards the end of the month, attention remained fixed upon the fate of Malaya and Singapore.

Since the beginning of 1940, Walter Walker's superiors had been trying to arrange a place for him at the Indian Army's Staff College at Quetta, Colonel Langlands writing a flattering letter of recommendation to General Sir John Coleridge, the Commander-in-Chief, Northern Command, at Rawalpindi. Although Walker was aged only twenty-seven Langlands wrote, 'My only excuse for making a special appeal for his name to be considered for the Staff College is because I am *convinced* that it would be *in the best interests of the Service* for him to go as soon as possible.'

But it was not until 1942, when the Japanese were cutting their way down the Malay peninsula towards Singapore, that the posting came through. Walker had served on the Frontier for three and a half years, far longer than he had expected. It might have seemed at the time that he had had to waste his military talents in a backwater of the currents that were scouring the world. It is unlikely that any officer in that year could have foreseen that the years in Waziristan would prove to be the best possible preparation for the ordeals to come.

The staff course at Quetta had been cut from a year to six months in order to mass-produce staff officers for the war against Japan. But, for this time, Walker could be with his wife in a married quarter and enjoy a little reminder of the social jollities of Quetta he remembered from the time before the earthquake.

Amazingly, at this time, training at the Staff College virtually ignored the war with the Japanese, concentrating on the Middle East. Many of the instructors had fought in the Western Desert and Syria and several wore the purple and white ribbon of the Military Cross. For Walker the course proved, after his experience as a staff captain, 'a very easy ride', as he put it. Yet at what was to be one of the most relaxed periods in his life, danger appeared from an unexpected quarter.

For about three years, Walker had suffered from occasional stomach pains. Being something of a Spartan, he had put this down to indigestion, or perhaps physical strain, and taken no action. The pain had become increasingly severe but, so long as it continued to be temporary, it was to be ignored. Then, at one of the Saturday night dances in Quetta, which would last until dawn, he collapsed in agony. He was rushed to hospital and an immediate exploratory operation was arranged. His appendix, it was found, was bent double.

Although illness kept him away from the Staff College for several weeks, Walker passed the course. But then, to his disappointment, he was not ordered east to fight the Japanese, or west to fight the Germans and Italians, but back to the North-West Frontier. He was appointed training major to the 3/8th Gurkhas at Loralai in Baluchistan, with orders to train them for desert warfare in the Middle East. After his experience at the Staff College this was a subject he was as well qualified to teach as any officer who had not fought there himself. But he was to theorise about desert warfare for only two months before, at long last, came the call to total war.

His appointment was to be as a General Staff Officer (Grade 3) on the staff of Lieutenant-General William Slim, who had commanded the 1 Burma Corps since 19th March, 1942. 'Burcorps', as it was known, comprised the 1 Burma Division of Burmese and Indian troops (now commanded by Major-General Bruce Scott, Walker's former Commanding Officer), 17th Indian Division and 7 Armoured Brigade. Since the Japanese had attacked across the Siamese border in January, and Rangoon had been evacuated, all that was generally known of Burcorps was that it was conducting a fighting retreat somewhere in Central Burma.

Walker's journey to war began with another long haul across India towards the frontier where the dangers now seemed to render those in the North-West insignificant. The railhead for reinforcements was deep inside Burma, at Mandalay, where the camps of reinforcement holding units had been established to await the daily road convoys south. As the train crossed the Burmese frontier, Walker imagined that he had timed his journey well and that he would arrive in time to catch the convoy to which he was allocated. Then, outside Mandalay, the train halted. Time passed and, fearing that the convoy might leave without him, Walker remonstrated with the guard; but the signals on the track clearly ordered the train to remain stopped. When, eventually, the train slowly dragged into the station Walker could see the reason for the delay. It was his first sight of modern warfare. Mandalay was burning; the train had been delayed while Japanese bombers flew over the city.

Devastation on this scale was not new to Walker, for it was minor compared with the total destruction of Quetta by the earthquake. But, while the disaster at Quetta and the fighting on the North-West Frontier had been adjacent problems they had remained separate in military thinking. In this new sort of warfare, the devastation of cities became not only an important strategic ingredient, but also tactical. This had already happened in Europe and South-East Asia and now it was happening on the frontier of India.

As he had expected, Walker had missed the convoy south and was sent to wait his turn in another in what was thought to be the comparative safety of Maymyo, the hill station for Mandalay. On arrival he found that the Japanese had been there, too, and that the Maymyo Club, where he was to stay, had been damaged by bombs. His stay, he was told, would last several days because Slim's infantry battalions urgently needed reinforcements to replace heavy battle casualties and these must have priority. So Walker impatiently paced the grounds and watched Chinese infantry training on the golf course, noting how raw they seemed compared with his Gurkhas.

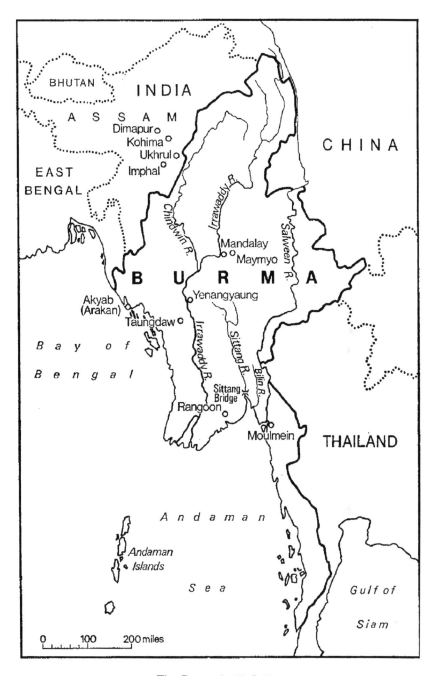

The Burma battlefields

At Maymyo, it was at last possible to hear in some detail of what had happened in Burma during the past three months. Short of having suffered complete annihilation, the Burma Army could hardly have undergone a more savage ordeal.

From the moment the Japanese High Command had decided that it would be necessary to occupy Burma early in their programme of conquest – to secure the right flank and rear of their drive south into Malaya and the Dutch East Indies – the outcome had been certain. The Japanese 15th Army, based on Indo-China, was allotted the task of first occupying Siam and securing the narrow Kra isthmus leading to Malaya and then invading southern Burma. Numbering in all some sixty thousand men, the 15th Army and the 5th Air Division, which arrived to support it, after the fall on Manila in the Philippines early in January, was a seasoned force. To its two divisions – the 33rd, which had been moved south from Nanking at the end of 1941, and the 55th – had been added the formidable Imperial Guards Division.

The Japanese infantry were of tough peasant stock and imbued with a ferocious patriotism, a contempt for their enemies and an aggressive self-confidence. The British awaiting them across the border had not, generally speaking, been trained in what was then known as 'bush warfare', which was often thought not to be proper soldiering. Moreover, General Sir Archibald Wavell, who had fought the Italians so brilliantly in Africa and had been appointed Supreme Commander of British, American, Dutch and Australian forces outside the Pacific area soon after the outbreak of war with Japan, was said to have a poor opinion of the Japanese as soldiers. This however was balanced by fearful stories of Japanese skill as jungle fighters and of their courage and inhuman cruelty to prisoners. Faced with such uncertainties, the modest British, Indian, Gurkha and Burmese force awaiting attack were particularly disheartened when they heard not only that reinforcements that had been intended for them were being sent to Malaya but that – to face what appeared to be the main threat to the south – much of their own air support was being sent there, too.

The defenders of southern Burma at this time amounted to two mixed divisions. Between the Siamese frontier and Rangoon stood the 17th Indian Infantry Division commanded by Major-General Jack Smyth, who had won the Victoria Cross in the First World War. Under him were five battalions of the Burma Rifles, five Gurkha battalions, a battalion apiece of Jats, Baluchs and Dogras, and a battalion of the King's Own Yorkshire Light Infantry. Farther north, astride such roads as there were running north from Rangoon through the Shan States, was General Scott's 1st Burma Division. The majority of infantry here also was from the Burma Rifles, with several battalions of Indians and a battalion of the Gloucestershire Regiment, as the main garrison of Rangoon.

On 15th January, Lieutenant-General Iida, commanding the Japanese 15th Army, launched his first major attack across the frontier, overwhelming a battalion of the Burma Rifles and capturing an airfield. His first principal objective was the port of Moulmein, beyond which lay the way to Rangoon, central Burma, Assam and finally India itself.

On 20th January, the Japanese 55th Division crossed the frontier and in ten days had fought their way without undue difficulty to the outskirts of Moulmein. The port was surrounded on three sides by tidal estuaries, easy to cross by boat and difficult to defend, and on the fourth by thick jungle. Smyth had estimated that two divisions would be necessary to hold Moulmein and two brigades to hold the twelve-mile inner perimeter defences but, as the enemy prepared to assault, all he could spare were four battalions of Burmese and Indian infantry and a few guns. After putting up little more than token resistance, this little garrison had withdrawn as best it could by boat across the mouth of the Salween river.

After this there were only two natural barriers between the Japanese and the capital: the Bilin and Sittang rivers. At the former, Smyth decided to make a stand on the west bank. As 17th Indian Division withdrew as fast as possible to this position, Wavell had telegraphed tartly, 'I do not know what considerations caused withdrawal behind the Bilin River without further fighting ... but bear in mind that continual withdrawal, as experience in Malaya showed, is most damaging to morale of troops, especially Indian troops. Time can often be gained as effectively and less expensively by bold counter-offensive. This is especially so against the Japanese.'

But, already, Smyth's division was in no position to mount a counter-offensive and all that could be hoped was that the Japanese could be held on the Bilin long enough for his weary troops to regroup and rest and even receive reinforcements. But it turned out that the choice of the Bilin as a defensive position was a mistake. Smyth's Gurkhas, Yorkshiremen and Jats fought bravely, but the Japanese, instead of attacking directly across the river along the line of the road and rail bridges, marched north, swung westward across the river and turned the defenders' flank. Now there was only the Sittang between the enemy and Rangoon.

The fighting retreat had, thus far, become increasingly dangerous but, now on the Sittang, it was to turn overnight into a disaster. The crossing to new positions on the west bank was to be made across a railway bridge for, without an elaborate boat service and command of the air, the river – deep, fast-flowing and as much as a thousand yards wide – was an effective barrier to any army. Smyth's men were exhausted when they reached the river. They had been fighting for three weeks and retreated one hundred miles. They were already confused by the Japanese technique of slipping through the jungle round their flank and setting up ambushes along the line of their retreat. Now their fear was that Japanese parachutists might seize the Sittang bridge before they could cross themselves.

The 17th Division began to cross the Sittang with the Japanese on their heels. As the first brigade reached the bridge, sappers hurriedly prepared for its demolition. At all costs the Japanese must not be allowed to take the bridge intact, so while these preparations were in progress the bulk of the division tried to form a bridgehead on the east bank. Then, in the early hours of 23rd February, reports reached Smyth at his headquarters on the west bank that enemy pressure was such that the bridge might be lost within the hour. Thinking that most of the division was safely across, Smyth gave the senior officer on the far bank his permission to destroy the bridge whenever he felt it was necessary. An hour later, the bridge over the Sittang was blown – but the bulk of the 17th Division was still on the east bank of the river.

The premature destruction of this bridge destroyed the 17th Division as a fighting formation, for less than three thousand, five hundred infantry had crossed the bridge, or were able to swim across afterwards, and of these less than half had been able to save their weapons. The way to Rangoon and beyond was open to the Japanese.

Smyth was now replaced by his senior staff officer, Brigadier D. T. ('Punch') Cowan and, a few days later, General Sir Harold Alexander arrived at Rangoon to take command of the Burma Army. But the capital was already doomed, evacuation was ordered and, on 8th March, it fell to the Japanese and the course of the campaign was drastically altered. Now the enemy had the advantages of holding the ports and could pour reinforcements and supplies into southern Burma at will in preparation for the drive northwards on central and northern Burma. General Iida was given as his objective a rapid advance with the object of bringing to battle the surviving British and Indian forces together with the Chinese armies which were training under the American Lieutenant-General J.B. ('Vinegar Joe') Stilwell in the region of Mandalay.

Singapore had fallen on 15th February, freeing more Japanese forces for the next phase in Burma. For the advance up the valleys of the Sittang and Irrawaddy rivers, General Iida now had four divisions – the 18th from Singapore and the 56th from Japan – and his air forces had been increased to a total of some two hundred and eighty-eight fighters and one hundred and twenty-six bombers and reconnaissance aircraft. Facing them were only remnants of the 17th

Division, the 1st Burma Division, together with the newly-arrived 7 Armoured Brigade, and two Chinese divisions, supported by only about one hundred and fifty aircraft. But Burma Corps was now given a new commander: the remarkable Lieutenant-General William Slim. Meanwhile, General Alexander, in charge of the overall strategy, had decided to concentrate Burcorps on the Irrawaddy route for the defence of the vast oilfields at Yenangyaung, three-fifths of the way from Rangoon to Mandalay. By the end of the month the stage was set for the next act.

At last there was a place in a south-bound convoy for Walker and he set out. Slim's headquarters had been moving north every few days with its light covering screen of infantry and, when Walker found it on 16th April, it was encamped to the north of the oilfields. As was the custom, the headquarters was spread out under trees – trestle tables, maps and notice boards, wireless trucks and hastily-dug slit trenches. There was intense activity because, as Walker was at once told, the Yenangyaung oilfields had been blown the day before and the Japanese were now making one of their encircling sweeps behind the troops retreating from the huge canopy of black oil smoke above the ruins.

This was no time for any exchange of gossip with new colleagues but there was one brief moment of gratified recognition. Walker recognised Major Brian Montgomery, the GSO (1), as an old and reliable friend. Montgomery, the brother of the future Field-Marshal Viscount Montgomery of Alamein, had been GSO (2) of Waziristan District Headquarters and they had often met during operations against the Fakir of Ipi.

A brilliant staff officer, Montgomery was something of a disciplinarian and punctilious to a degree: Walker never knew him draw a line on a message pad without using a ruler. He was brisk, sometimes finicky but, Walker knew, highly professional and unlikely to become flustered in a crisis. For his part, Montgomery was as pleased to see Walker, in whom he also recognised an efficient professional and a disciplinarian. But there was no time for reminiscence. Montgomery told Walker that that same night he would start work as duty staff officer under instruction. So, knowing nothing of the tactical situation, Walker found himself helping to mount a counter-attack against the Japanese who had set up a road block between corps headquarters and the troops retreating from the oilfields. Somehow in the confusion and darkness this was successfully achieved.

Corps headquarters was necessarily small – no larger than the customary brigade headquarters – and there was little equipment that could not be packed into a column of Jeeps. It was to be Walker's task to reconnoitre the next site for the headquarters and lay it out. Thus, every two days or so, he would drive north up the road or track to look for a suitable place that offered concealment from the air, a natural defensive position and the best possible overland communications. Occasionally he chose a village but these were obvious targets and the Japanese Zero fighters, which dominated the air, regularly struck at the corps' line of retreat. More often the site would be in a patch of jungle or under coconut palms. It would be dangerous to choose an isolated clump of trees to which Jeep-tracks could be seen to lead, so Walker tried to put himself in the position of a Zero pilot looking for the headquarters and then choosing the least likely site.

Usually he chose well – each new site being some twenty miles to the north of the last – but once there was an alarm when the Japanese, using river craft on the Chindwin, outflanked Corps Headquarters itself and a counter-attack had to scramble into action to force them back. In following this hectic routine, Walker found that his long experience in the very different terrain of the North-West Frontier proved its value. Then, too, he had belonged to a light-scale, highly-mobile headquarters, always under threat of attack by a skilled and courageous enemy. Both he and Montgomery had had this grounding but so, too, had the corps commander. Slim had also been on the Frontier and, more recently, had commanded a Gurkha battalion in the Middle East and, later, a division during the last year's operations in Iraq. Tough and imperturbable, Slim quietly dominated his corps, radiating a confidence that he cannot have felt himself, although, even at this stage, he had not given up all hope of catching the Japanese off balance and delivering a counter-blow that would drive them south. Working day and night shifts of eight hours, Walker saw much of Slim and later only remembered one occasion when he showed strong emotion.

General Scott's division had been encircled and was hoping to break out in a night move. The complex orders for this, and for various covering and diversionary moves by other formations, had been passing through Walker's hands and copies were being circulated in the usual manner. A major action seemed likely and Slim was prowling the headquarters site, sunk in thought. Then Walker saw him look up at the board on which were pinned copies of the orders he had circulated. Suddenly Slim frowned in anger and, striding over to Walker's table, stormed at him. He had, said Slim, omitted to include the Director of Medical Services among those due to receive copies and that, with a battle imminent, was unforgivable. Walker never forgot it again.

Slim's own superior, General Alexander, was directing the strategy of the campaign from his headquarters near Mandalay but Walker occasionally passed him on the road and would be stopped and asked for the latest news. In contrast to Slim's relaxed appearance, Alexander always wore his high-crowned Guards' cap and a polished Sam Browne belt. But, like Slim, he never appeared ruffled, however disastrous the outlook.

By 25th April, Burma Corps found itself making a last stand before Mandalay itself and Slim had had to abandon all hope of a counter-offensive. The Chinese armies on the left flank had suffered a series of shattering defeats and the two British and Indian divisions had been terribly mauled, the 17th Division, despite the arrival of reinforcements, being reduced to a loose-knit collection of exhausted survivors. It seemed likely that Burma would have to be evacuated completely and

Tom Pocock

Slim planned that the Chinese should fall back into China – perhaps together with some British troops – and the rest into India.

The last practicable line of defence within Burma was the Irrawaddy. Mandalay, on the east bank, had to be evacuated by the end of the month and what was left of Burma Corps withdrew across the road and rail bridge over the river and, on 30th April at one minute before midnight, this was blown. The explosion announced the loss of Burma to Japan.

During the long retreat – some one thousand miles in about three and a half months – the Burma Army had never disintegrated. It had had to fall back constantly not only in the face of fierce Japanese attacks and encirclements but in the hope of reaching India or some stable defensive positions before the monsoon broke in May. The final act was the crossing of the Chindwin and there the last of the heavy weapons and vehicles were destroyed or abandoned. Walker himself lost most of his kit when his Burmese batman deserted to the Japanese, taking with him Beryl's wedding present: a pigskin case fitted with ebony-backed brushes.

The first monsoon rains fell on 12th May but, by then, Burma Corps was straggling into India. Walker, watching them, was glad to see a Gurkha battalion still marching in military formation and each man still carrying his rifle slung. Corps headquarters had suffered badly. They had long lacked protection against malarial mosquitoes: no nets, no mepachrine, no protective oil. When Slim's staff crossed into India only half a dozen officers were still on their feet. On 20th May it was all over, officially, when General Alexander's now non-existent command was formally disbanded and the surviving troops came under command of an Indian Army corps.

The Burma Army had suffered a total of 13,463 British, Indian, Gurkha and Burmese casualties, against 4,597 Japanese killed and wounded. It had lost virtually all its weapons – except personal arms – ammunition, transport and supplies. In the air, losses had been almost exactly equal, with just over one hundred aircraft lost by each side.

65

So the first Burma campaign ended in total victory for Japan. Not only was the country itself occupied, but the Japanese had succeeded in cutting the vital Burma Road to China so that now the only supplies from the outside world reaching that country had to be lifted in aircraft flying 'over the hump' of dangerously high mountains. With success throughout the Pacific and South-East Asia complete, the enemy could now prepare for the invasion of India itself.

During these months of battle, Walker had not himself commanded troops in action. But any tactical experience he would have obtained in this way was more than compensated for by the overall picture of the campaign which passed before his eyes each day in the form of signals. There would be time later to learn tactical lessons, this had been a time for understanding strategic essentials. It was clear to him that Burma Corps had not had a chance. Most of it had been trained for desert warfare in the Middle East and had had to pick up whatever it could learn from the Japanese about jungle fighting in horribly practical lessons. There had been delays, too, in bringing the large Chinese reinforcements into battle. Finally, one positive lesson had been in the importance of the mobility and firepower of 7 Armoured Brigade, which had been able to switch quickly from one vulnerable point to another, so frequently averting disaster.

Once safely inside India, the survivors of Burcorps faced a humiliating shock. For many weeks they had kept up their spirits by thoughts of the security and comfort that awaited them over the frontier. They imagined passing through powerful forces of the Indian Army, poised to protect them and to launch a massive counter-offensive against their pursuers. Meanwhile they themselves, they thought, would be housed in clean, dry transit-camps to await leave. This was not to be the case. Arriving at the frontier garrison town of Ranchi they were appalled to find the hospitals filled to overflowing and the comfortable transit-camps non-existent. Those able to march – or, rather, to walk – were directed away from the town and told to make their own bivouacs just as they had in Burma. And, as for the mighty, avenging Indian Army, that was nowhere to be seen.

At Ranchi, Walker was asked by Montgomery to write the Corps' war diary. Meanwhile he was one of those granted leave and, soon after, began the week-long rail journey back across India to Quetta for a month with Beryl. There, just as the veterans of the Libyan desert had once been the talk of the Staff College, now it was this thin, sallow-skinned staff officer from Burma. Walker was asked to lecture on his experiences and did so. But that evening an instructor from the Staff College came up to Walker as he dined with Beryl at the Quetta Club. Both staff and students wanted to hear more, he said. Particularly important to them was Walker's account of conducting a campaign from a small, highly-mobile corps headquarters. When the time came for a return to Burma these would be needed again, these streamlined headquarters able to move quickly by Jeep, or even aircraft. Would Walker therefore lecture again? Protests that he was still tired and was meant to be resting had to be dropped and Walker found himself indoctrinating the future officers of the avenging 14th Army with the ways of jungle war.

The Burma battlefield. A Gurkha soldier rests beside a Japanese hillside bunker that has been taken by assault

At the start-line Indian infantry awaits the order to advance on Mandalay in
March 1945

The Malayan battlefield. A Gurkha rifleman alert
in the damp gloom of dense jungle

The respite could not be long and soon Walker received orders to return to Ranchi as Deputy Quartermaster-General of a division preparing for the return to Burma. Shortly before he was due to leave, there was a change of plan and he was asked to remain at Quetta as an instructor at the Staff College to lecture on staff duties. Again he found himself directly junior to Montgomery, who was in charge of his division at the College, and was delighted to find that the new Commandant was Brigadier Geoffrey Evans, who had won the Distinguished Service Order in Eritrea.

Now Walker could continue to transmit his experiences in Burma to eager young officers preparing for the counter-offensive. He told them how a compact, mobile headquarters should be composed and used and how to write the orders, appreciations, papers and signals that would be their lot as staff officers. Not only was Evans trying to recruit as many instructors as possible from those with recent experience of operations but he arranged for as many lectures as possible from visitors fresh from fighting in the Middle East, or on the frontier of Burma.

The Walkers could now enjoy some domesticity and the couple were grateful that chance had brought them together for the birth of their first child. One morning, while Walker was lecturing to his class, he saw the adjutant peer through the window, grin and give what he took to be a rude gesture with two fingers of his right hand. Walker returned the signal but the adjutant shook his head and called out, 'No, no! It's twins!'

The students cheered, the Commandant, who had heard the uproar, asked what it was all about and Walter Walker leapt on to his bicycle to ride over to the hospital where he saw Beryl and their twin sons. The peaceful interlude had become idyllic.

3

Burma re-Conquered

The most satisfying experience in the career of a professional infantry soldier is said to be the command of a battalion, particularly if it belongs to his own regiment. However senior he may later become, he will never again exercise his powers of leadership with such direct intimacy; never again will he impress his own character on others with such immediacy; never again will he come to know something of the personality of almost every man under his command.

That this supreme moment was fast approaching for Walter Walker became apparent when, after instructing for three six-month courses at the Staff College, the long-awaited orders for active service arrived. The fourth battalion of the 8th Gurkhas had suffered heavy losses fighting against the Japanese on the Arakan coast of Burma and Walker was to join them as second-in-command.

Stories of the Arakan fighting that had reached Quetta were generally confused, and with good reason, because the situation itself was one of confusion. The Burmese shore of the Bay of Bengal was of obvious strategic importance because it seemed to offer the only means of turning the flank of the Japanese forces assembling for the invasion of India. Thus, since its evacuation in April, 1942, General Slim, who was appointed to command 15 Corps in Assam, had been hankering after a return. Now that some intelligence was filtering through from Burma, it appeared that Japanese strength there was not as formidable as it had earlier seemed. It amounted to no more than four divisions spread over an enormous area: one in the north-east facing the Chinese; two on the Assam front and Central Burma; and one – the 55th – in Western Burma with little more than a brigade group in the Arakan. It seemed that an attack there – either a straight overland offensive, an amphibious landing, or a long-range penetration expedition, or a combination of all these – might well succeed.

The attempt, which was a combination of these three possibilities, was made; controlled not, in the event, by Slim but by General Sir Archibald Wavell, the Commander-in-Chief of the Army in India. It was a failure. The division with the task of mounting a direct offensive was not trained, or equipped, for its task and crumbled in the face of violent Japanese counter-attacks. Beginning with high hopes and much publicity about driving the enemy out of Burma at the end of 1942, the offensive failed to reach its principal objective, the port of Akyab, early in the following year, and, by May, had been driven back to the positions from which it had started.

Slim was determined to try himself and, having learned the lessons of the first Arakan campaign, became convinced that a more ambitious offensive on a wider front could succeed. Therefore, in November, the whole of 15 Corps prepared to move down the coast towards Akyab. Among the infantry battalions in the 89 Indian Infantry Brigade of the 7th Indian Division was the 4/8th Gurkhas.

Initially, the second Arakan offensive followed the same course as the first. The successful British advance was met by a ferocious Japanese counter-offensive. But this time the British held. The fighting, across low hills covered with dense bamboo thickets, and across dry river-beds – known as *chaungs* – was intense and at close quarters. The British and the Indians, Gurkhas and Africans, fighting alongside them, still tended to think of the enemy as being invincible. But they now also knew what the Japanese did to their prisoners, so that no quarter was expected or given, so giving the fighting an extra edge of ferocity.

Time and again the British found themselves outflanked, cut off or surrounded, but they had come to accept this situation as inevitable and formed themselves into 'boxes', a formation directly derived from the 'squares' of Waterloo. Whole base areas could be contained within a defensive box and, while the enemy squandered his infantry in desperate assaults on the perimeter, the defenders could rely on supplies dropped by air and hold on.

The hardest-fought of such battles was that for 'The Administrative Box', headquarters of the 7th Division itself. With the divisional commander, Major-General Frank Messervy, fighting alongside his staff and clerks, the attacks were held and among the troops defending the perimeter were the 4/8th Gurkhas. They, like others, suffered much both before and after the box was finally relieved on 23rd February, 1944. Battle casualties were expected to be heavy in fighting like this but there were also atrocities to be suffered. One of the young British company commanders in the battalion was wounded and sent safely back to the field hospital. This was overrun by the Japanese, who murdered the wounded where they lay, then lined up the doctors and orderlies and shot them.

Less than a month after this battle had been decided, the battle-weary battalion suffered a shattering blow. The Commanding Officer, Colonel Twiss, and his adjutant, were travelling back to Brigade Headquarters in a tracked carrier when it struck a massive land-mine. Both were killed instantly. As a temporary measure Major Derek Horsford, of the 1/8th Gurkhas, who had been with Walter Walker on the North-West Frontier, was put in temporary command while a new CO, Lieutenant-Colonel N. D. ('Pooh') Wingrove was summoned from India.

Soon after this, Walker found himself at Dum-Dum airport, Calcutta, on his way to the Arakan. It was night and nobody seemed to know when, or whether, an aircraft would be flying in the direction he followed. At the control tower he ran into another British officer wearing a wide-brimmed Gurkha hat. 'Aha, a fellow Mongolian!' they greeted each other, then introducing themselves, one as the new CO, the other as the new second-in-command of the 4/8th. So it was that these two Regular Army officers arrived to take over the survivors of this war-raised battalion of their regiment, officered almost entirely – with the exception of Horsford, the temporary CO – by ECO's: 'Emergency Commissioned Officers'.

The moment of change in command is always sensitive but this occasion bristled with potential problems. Two officers, one without experience of fighting the Japanese, were to take over a battalion that had been blooded in battle to the fullest extent possible if it were still to remain a fighting battalion. Both at divisional and brigade headquarters, Wingrove and Walker were warned that the 4/8th had been badly shaken. Clearly they would need careful handling.

Colonel Wingrove arrived at battalion headquarters and formally took over command from Major Horsford on 5th April.

The beginning was not auspicious. Walker, immaculately dressed, well-groomed and fit, was inspecting the battalion as it went through its daily routine in reserve. He was not particularly impressed. They had, of course, been through a prolonged ordeal by battle, but, to his eyes, there was a certain lassitude in discipline, a need for alertness and vigour. This impression was not always easy to define, as he walked through the camp. But then he came upon an example, minor in itself, but none the less significant. A company of Gurkhas was on physical training parade under the command of a Gurkha officer, whom Walker recognised, but there was no British officer in sight. Questioned, the Gurkha officer explained that his company commander must have been unexpectedly delayed.

At that moment, Walker saw strolling down a path towards him a dark-haired young officer. He proved to be the missing company commander. Walker coolly asked his name and was told it was Myers. He was then asked his name. Walker curtly said that he was the new second-in-command and that he had already noticed some shortcomings in the battalion. Why, for example, was Myers ten minutes late for his company's physical training parade? In future Myers would, he announced, conform to the standards of the regiment. As he said this, he noticed a cynical, bored expression on Myers' face, which seemed to say, 'Oh, my God, the new broom has arrived!'

Peter Myers was not alone in his resentment of Walker. 'The new broom' was, he thought, altogether too cocky and too much of a martinet. They – the officers of the battalion – had been through the furnace of war at its most savage and Walker, so far as they knew, had not. He therefore had no right to bully them like a bunch of recruits.

Their vague dislike and distrust in these early days came to a head over a relatively trivial incident. One of the duties of the Intelligence Officer, young Lieutenant Peter Wickham, was to wake Wingrove and Walker before dawn each morning for a tour of the battalion positions on the perimeter at stand-to. One morning Wickham himself was not woken by his orderly, but the colonel and second-in-command had themselves woken and walked across to see what had happened to their Intelligence Officer. He was still asleep but, rather than wake him themselves, Walker recommended that he should be left as he was. The two then carried out their inspection and, when Wickham came rushing after them, full of apologies, Walker severely disciplined him for failing not only to wake them but to accompany them on their tour. To Wickham's brother officers, to whom a jog on the shoulder from Walker would have seemed the obviously sensible and understanding answer to this little problem, this seemed harsh, bull-headed behaviour.

Walker now decided that his inspections must become increasingly close and careful and had taken to riding round the camp on a charger that he had found in the mule lines. He was also in charge of training so that his trim, impeccable figure sitting straight-backed on his horse became a symbol of ruthless and demanding authority to a battalion that craved above all rest and relaxation. There was grumbling in the officers' mess and no doubt that the first impression that Major Walker had made was that of a martinet.

Then there was the contrast with Horsford. He had only been in command for a month but, as Walker had been told at divisional headquarters, he had been a brilliant success. Tall, wiry and a natural leader of men, he was also a quick-witted tactical improviser and capable of ruthlessness tempered by kindliness and humour. When he had first arrived, he found the battalion severely shaken and slack in some ways, so his first move was to send ten Gurkha officers back to Quetta. This decisive act provoked rage at the depot and the senior officer, whose permission for such drastic action should have been sought, wrote to Horsford telling him that he was under arrest. This letter provoked laughter on the battlefield.

Meanwhile there were increasing signs that the fighting in the Arakan was about to become little more than a sideshow to a main event. Throughout 1943, the Japanese had had only four divisions and supporting troops in Burma. Then a fifth had arrived from Java for service in the Arakan, a sixth had arrived from Malaya, a seventh was coming in from Siam and, it was reported, other major formations would be following from the Pacific. Thus, early in 1944, the Japanese had assembled two armies in Burma – one of three divisions, the other of four – and this could only mean that another major offensive was about to be launched.

This state of affairs was most satisfactory from the Allies' strategical point of view for it meant that large numbers of seasoned Japanese soldiers were occupied in Burma and unable to threaten, say, Australia. As Japanese intentions became increasingly clear, the assessment of them suited General Slim remarkably well. In their brilliant campaign of conquest, the Japanese had seriously over-extended themselves and, in particular, their communications. Their garrisons on Pacific islands, in the Philippines, the East Indies, Malaya, Burma and China were now under increasing pressure and were almost wholly dependent upon their sea communications with Japan. Their supply ships were being sunk – mostly by American submarines – at a mounting rate which, if it continued – as seemed inevitable – would result in the strangulation of their far-spread forces. Therefore a quick and decisive victory was essential to Japan.

This, it seemed, could only be achieved on the Burma front. If they could destroy the 14th Army, they would not only have India and its limitless resources at their mercy but would finally cut China off from all outside help. Therefore, it seemed that, sometime in the spring of 1944, the Japanese would attempt a knock-out offensive against the defenders of Assam.

Slim, also, was hoping for a decisive battle in Assam. His intention was, of course, to recapture Burma but the terrain he would have to cross was in itself a natural defence and, combined with the tenacity of its Japanese defenders, could become impregnable or, at least, demanding an unacceptable cost in the lives of the attackers. So it would be necessary to destroy the enemy before plunging into the mountains and jungles of Burma and, if that was to be done, the ideal killing ground would be the plain of Imphal. This, strategically, could be looked upon either as the Gateway to India or the Gateway to Burma. With each side taking an opposing view, a collision of aims and armies became inevitable to the satisfaction of both.

Slim planned to concentrate his 4 Corps in the Imphal area to take the shock of the Japanese attack, while leaving 15 Corps on the Arakan coast to pin down as many enemy as possible and to act, if necessary, as a reserve for the Imphal battle. He expected that the Japanese would try to hold the Arakan front and the northern front, where the Chinese were pressing on them, while throwing their main blow in the centre. This, he forecast, would fall on 6th March, or thereabouts, and would be led by the four divisions of the Japanese 15th Army. He was right. The Japanese offensive began on 6th March.

The ferocity and skill of the assault surprised even Slim. Using their familiar tactics of infiltration and outflanking, the Japanese seemed invincible as they tore down the defences of 4 Corps. The key positions for both sides were the small towns of Imphal and Kohima, some sixty miles to the north, in the hills and covering the strategic railhead at Dimapur. After the first week of fighting it looked as though both might be overwhelmed. Large-scale reinforcements had to arrive soon and, in an airlift of remarkable dimensions, the 5th Indian Division was flown from the Arakan straight into battle on the plain of Imphal.

Among these reinforcements was the 4/8th Gurkhas. Their journey in battle was far less simple than the massive airlifting of troops that was later to become commonplace. It began with a nine-day journey by road, rail and steamer to an airfield at Sylhet in Bengal from which the 'fly-in' was being mounted. They arrived safely and, on 9th May, took over a sector of the defences in the hills, twelve miles to the north of Imphal and just to the east of the road from Imphal to Kohima.

Their first operation was to take part in an attack on one of the ridges dominating the road. Artillery support was sparse, the climb was exhausting and the enemy was stronger and more deeply dug into the hillside than had been expected. The attack failed, after costing the battalion eleven dead – including a British officer – and thirty wounded.

The task of the battalion and other formations of the 5th Division, to which it now belonged, was to clear the road to Kohima, initially advancing some thirty-five miles northward as far as Ukhrul. It was not so much the road itself that was to be the battlefield but the hills which commanded it. Now there was to be some ugly fighting. The Gurkhas struggled up steep hillsides, slippery from torrential monsoon rain, in the face of point-blank machine-gun and artillery fire. The Japanese were waiting in bunkers of such strength that conventional artillery support and air strikes were unable to dislodge them.

One assault after another went up the sharp ridges, now stripped of any cover from trees and scrub by bombardment, and each finally came to a halt before the subterranean defences. Between attacks, the battalion clung on to its positions under continual fire. In easier terrain, the answer would have been to bring up tanks but here it seemed impossible. When it seemed that, after a series of enemy positions had been carried at heavy cost, the final defences would prove impregnable, it was decided to try to get tanks into the battle.

A troop of three American-made Stuart tanks were brought to the foot of the hill, then with heavy tackle fastened to tree-stumps, slowly winched up the steep slopes. Finally, two were ready to move forward with the infantry, the third being held back to give longer-range supporting fire. As each bunker was approached, the tank would bombard it with high-explosive shells and rake the loopholes with machine-gun fire, while the infantry followed. Then, when almost on top of the bunker, the gunner would switch to solid armour-piercing shot so that the Gurkhas could rush the final yards without being themselves blown up. The crescendo came when, shouting Gurkhali war-cries, the infantry ran past the tank, threw grenades through the bunker's loopholes and, when they exploded, scrambled inside with drawn kukris or fixed bayonets. And so the hills were captured.

Now the 4/8th were given a short respite in reserve, held in readiness for the coming battle of Ukhrul, an important Japanese communications centre. At the beginning of July they marched into the battle area, the stench of unburied enemy dead greeting them. It was again to be a stiff fight, with Japanese artillery firing over open sights at point-blank range, so that the report of the gun was heard after the explosion of its shell. Two brigades were used to capture Ukhrul and the 4/8th were the spearhead, sometimes seizing and holding a position only forty yards from the enemy, the entire battalion packed into an area seven hundred yards wide and one hundred deep. Bad weather stopped air-drops of supplies for two days and the Gurkhas, without rations for that time, showed signs of exhaustion. But the enemy were even more exhausted and when the second brigade used their own favourite tactic of encirclement and attack in the rear, they broke and the survivors abandoned the town. With the fall of Ukhrul the road was open and Slim could contemplate turning the tables on the enemy with a counter-offensive into Burma.

Now that the Japanese invading armies had been broken and their rearguards chased back into the hills, there was time for rest and recuperation. Few units of the 7th Division needed this more than the 4/8th Gurkhas, but for them it was to be revitalising and retraining. When they reached the site of their camp on a steep hillside a few miles to the north of Kohima, they were haggard after the months of battle, and reinforcements could be instantly identified because their ribs did not ridge their chests. Now they were issued with double rations and ordered to construct their new camp.

The battalion was in these beautiful hills for five months but, while neighbouring battalions enjoyed rest and recreation, the 4/8th found that Walker had other plans for them. For the first four months, he was second-in-command and training officer and then, in November, succeeded Wingrove as Commanding Officer.

The battalion must not only become more skilful and tougher fighters than the Japanese, he had decided, but know themselves to be so. Until this was achieved, rest and recreation were last on Walker's list of priorities. The battalion's Standing Orders, which he now drew up, began with a long list of 'Slogans of 8 Gurkha Rifles'. These were simple and savage and every man in the battalion had to learn them by heart.

'It is the ambition of every man of the 8th GR to redden his Kukri and bayonet in the enemy's blood.'

'DESTROY the enemy, NEVER allow him to retire.'

'Move quickly, fight fiercely, shoot low.'

'It is the last ounce of TOUGHNESS that wins battles; so I must be tough, look tough and act tough.'

'Shoot to kill – kill to live.'

'I prefer life to death, therefore I must never surrender.'

'To surrender to the enemy is shameful. Even if all my comrades are killed, I must fight the enemy single-handed and when I have no more ammunition left I must attack and kill the Japanese with my kukri and bayonet.'

And the motto of the battalion was:

'LIVE HARD, FIGHT HARD, and when necessary, DIE HARD.'

The surrounding hills were riddled with old Japanese bunkers, tunnels and trenches and Walker made the battalion study their construction and then build their own defences to the same type of design. They must expect to be encircled and regard this state as normal. Therefore defences must always cover one hundred and eighty degrees. Major Peter Myers, the most naturally soldierly of the battalion's Emergency Commissioned Officers, who had won the Military Cross at Ukhrul, was ordered to plan such a position and construct it with his company. They worked all day, then in the evening Walker arrived to inspect their work. He was not satisfied. The company would have to stay in the position all night and dig it again at first light next day.

Walker was punctilious over detail. Slit trenches had to be four and a half feet deep. Barbed wire must be strung between the stakes at three different heights so that the enemy could not even wriggle beneath it. There must be a constant supply of sharpened *panji* stakes – each man would carry a bundle – so that these could be planted instantly round any defensive position

When satisfactory defences had been built, the companies took turns in attacking and defending them. In night tactical exercises, Walker used all the Japanese tricks – notably the initially terrifying 'jittering' of positions with screams, grenades, yelled taunts and bursts of tracer – and light machine-guns fired live ammunition on fixed lines.

Walker trained and then commanded his battalion through his British officers. Unlike Horsford, he had not got the easy, relaxed touch in his daily professional relations. Nor was his Gurkhali as fluent as it might be. Therefore he stood as a paragon of military skills and virtues, insisting that his officers emulate him and that they instruct their men likewise. Yet he was an even more stern disciplinarian with himself than with his young officers, all of whom were in their twenties, usually early twenties. If there were doubts whether a sentry could remain alert for six hours at night in the jungle, Walker would experiment himself. If it was doubted whether a rifleman could carry a fifty-pound pack on a forty-mile march across the grain of hill-country, Walker would himself make this march. He only slept for between two and four hours at night.

Officers who only once strayed even fractionally from his requirements never strayed again and remained with the battalion. One of his young officers, Patrick Davis, was to remember, 'Walter Walker drove us hard.... He caused much heart-ache, much grumbling, sometimes feelings for which grumbling is a polite euphemism.... Occasionally, we thought him unjust, but only occasionally. For myself, I was in awe of him, but committed to his way of running things. I thought it right that he was tough and hard on us. We were not playing a game. There was no umpire to rule the enemy offside. No one would bring back to life our dead.'[1]

Another, Denis Sheil-Small, who remembered Walker as a terrifyingly smart and professional officer at the Quetta training centre, ruthlessly recording any error in a little black notebook, had similar feelings. Once he and several others were slightly late for a conference at battalion headquarters. Their punishment was that, for three nights, each officer would make the rounds of all four companies and the headquarters and sign an exercise book placed at each for the purpose. On his return to the officers' lines, he would wake the next victim, who would do likewise, until the first of them was woken again, and again. So it went on until the practice stand-to at first light. None of them was late again.

This nocturnal stumbling through the forest and among slit trenches had another purpose. It helped these young men to feel easier in the dark. The Japanese had been masters of jungle night-fighting and soon they would have to be mastered in this themselves. Walker insisted that, even when occupying fortifications, the battalion must think aggressively; his doctrine of 'offensive defence', stressed the importance of ambushes.

Many nights were spent lying in ambush, sometimes staying in position without moving for forty-eight hours, the only communication between the hidden men being vine tendrils, tied to wrists which could convey simple signals by gentle tugs. Experience had shown that Japanese reaction to an ambush was swift but predictable. When the ambush was sprung, any surviving soldiers in the lead would dive off the track to outflank and surprise the ambush party and a mortar, ready for this some thousand yards back, would bombard the ambush position. Therefore the ambush would have to have its own all-round protection by 'stops' of riflemen covering the flanks and rear and a party could be in readiness for a wide encircling movement to knock out the mortar. These, and many other such tricks of the profession, were rehearsed until they became reflex action.

The war itself had long disappeared over the hills, far away into Burma, but the 4/8th lived day and night with war, this filling their thoughts and dominating their lives. Digging defences, bayoneting dummies, navigating patrols across strange country by day and night, planning raids and ambushes, firing rifles, machine-guns and mortars on the ranges, checking weapons, rations and equipment, all the routines of war were becoming their way of life and nothing outside war had any relevance to them.

Hard taskmaster that he was, the professionalism in Walker's character was balanced by integrity, generosity and warmth. His officers had come to realise that he was utterly 'straight', that he never made excuses for himself, just as he never accepted them from others, and that he was a man of his word. This warmth showed not only when relaxing in the mess but when giving orders, in a polite and friendly manner that implied that he knew there was no need to emphasise them aggressively. Once his officers understood his ways and could meet his standards, they recognised a high quality of leadership.

On 20th December, 1944, when Walker had been in command for nearly a month, they were ready. On that morning, he issued an Order of the Day beginning, 'The Battalion is about to proceed on active operations against the Japanese.

'The fighting in front of us will be tough and we shall have to exert every ounce of our strength. But we shall defeat the enemy just as we have defeated him before, and just as other Gurkha battalions have defeated not only the Japanese but the Germans also....

'I know we shall kill and utterly destroy the Japanese whenever and wherever we meet him. As Commandant of this Battalion I pin my faith in you. You have character; you have grit and guts; you have supreme courage; you have high morale; you have discipline; you have pride in your race; and now that you have confidence as a result of our recent training you are all going to be an everlasting credit to the cause which roused the manhood of Nepal and the land which gave you birth.'

He was right. The battalion had all these things and, to those who had known it five months before, the transformation was complete. Then it was war-weary, physically drained and would, had it been a more sophisticated body of men, have been cynical. Now it was trained to a high pitch, physically as strong and healthy as it could ever be and brimming with aggression and high spirits. That some – perhaps many – would not survive the coming ordeal was a thought submerged beneath the zest of youth.

The first task that General Slim had set the 4/8th was an inspiring one. They were to use the enemy's tactics against him and do so on the grand scale. When it became evident that the Japanese were not planning to make their stand to the west of the Irrawaddy but would use the river as their principal defence, he conceived a bold and original plan. While the main weight of the 14th Army would advance eastwards, two divisions – the 7th and 19th – would encircle the Japanese armies in the south.

While 33 Corps, with an extra division, would force a crossing of the Irrawaddy north and west of Mandalay, so drawing more Japanese divisions to its defence, 4 Corps would strike south in a three-pronged advance, cross the river and at the climax of a wide, encircling sweep, destroy the enemy's bases and communications centres and cut their escape route to Rangoon and Siam. Thus, 33 Corps would be the hammer, 4 Corps the anvil.

This motif was to be repeated within 4 Corps. One brigade, relying solely on pack transport and air-dropped supplies, would carry out a wide sweep through the jungle to cut off any Japanese forces than might impede the main advance. This was to be 89 Brigade and the leading battalion was to be the 4/8th Gurkhas. This place of honour could only be given to the best of battalions and seemed to Walker and his officers their reward for their unrestful recuperation at Kohima.

The essence of the brigade's advance was secrecy for, at the end of it, the Irrawaddy had to be crossed and a bridgehead established on the far bank. At the chosen point the crossing would have to be one of more than two thousand yards, the longest river-crossing ever attempted. Obviously, if the Japanese knew the exact point, the result could be disastrous.

The main advance south began on 19th January and, as hoped, 89 Brigade made its sweep undetected, seized the village of Pauk and had to skirmish to clear some high ground fifteen miles farther on. From there the distant river could be seen. The enemy must have known that a large force was churning through the jungle on the west bank and that a crossing might be attempted but the point chosen for the assault remained secret. It could, indeed, have been anywhere along a fifty-mile stretch of river and, for such a front, Japanese forces were limited.

The night chosen for the 7th Division's first major crossing of the Irrawaddy was that of 13th/14th February. In their brigade sector, the 4/8th Gurkhas had no immediate crossing role – that was for a battalion of the South Lancashire Regiment – but were to help manhandle the assault boats down to the river bank. One of their officers, Myers, was, however, to accompany the first wave and become beachmaster on the far bank. The leading assault company paddled safely across in the dark and dug in, to await the arrival of the main body.

The first flight of assault craft set off just before dawn, in the hope that their infantry could be ashore and dug in by first light. But of all military operations, the assault river crossing is among the most difficult and liable to sudden confusion. This was one such. There was muddle and delay in the embarkation and, once out in the swift, choppy stream, some boats' engines failed and others sprang leaks; finally, the carefully arranged pattern of the assault became hopelessly confused. At this moment, the Japanese opened fire from the far bank, immediately killing, among others, two company commanders and sinking the commanding officer's boat. The survivors turned unsteadily for the west bank as the dawn came up over the Irrawaddy. One company of the South Lancashires had landed and was cut off among the enemy on the far bank.

A second attempt had to be made but, now, in full daylight. Later in the morning a Punjabi battalion managed to cross safely for, to everybody's surprise, the enemy had not only failed to exploit the earlier disaster but had actually withdrawn to regroup. With more than a battalion to hold the bridgehead, the crossing was safe and the 4/8th Gurkhas – led by an immaculately turned-out Walker with his green neck-cloth tied, someone remembered, like an eighteenth-century cravat – made their own crossing without incident. On the east they were allotted a sector of the expanded perimeter to hold

The expected Japanese counter-attack did not materialise and did not seem likely to do so in the immediate future. The 4/8th now patrolled intensively, far more actively, so it seemed to them, than the other battalions on the perimeter. Indeed, it sometimes seemed that their patrols, criss-crossing each other across the plain in front, were not entirely necessary. This may have been partly true, but Walker was using this activity to flex their military muscles and prepare them for the contact which could not be delayed much longer.

The country they now faced was in contrast to the jungle-covered hills they had crossed on the journey from Kohima. Here, across the Irrawaddy, was the central plain of Burma. Flat and sandy and seamed by dried watercourses called *chaungs*, the plain was dotted with villages each sprouting with palm trees, cactus and a pagoda. These villages, looking so quiet and serene in the distance, were to dominate their lives in the weeks to come.

The 4/8th Gurkhas' patrols, which might be as strong as two platoons and stay out for three days and nights, probed in search of Japanese positions. Only the most detailed reports would satisfy Walker. It was not enough to tell him that it appeared as though the enemy held a village in strength and had dug bunkers beneath the surrounding cactus hedges. He would want to know how many machine-guns were mounted there.

Behind the bridgehead perimeter more crossings had been made in preparation for the deep plunge across the Japanese lines of communication which Slim hoped would be the decisive stroke of the campaign. Here on the central plain he would at last be able to use his armour and his overwhelming air power to its full extent, no longer hampered by jungle and razor-backed ridges and the drenching monsoon.

Soon after, the excitement of the secret march to the Irrawaddy and the tension of the crossing itself all seemed worthwhile. A patrol of the 4/8th had climbed a ridge and were using it to watch for Japanese movement on the plain below. Then, one morning, the plain was suddenly seething with tanks and armoured cars, artillery and ammunition convoys. One of the main eastward thrusts of the 14th Army was passing through their bridgehead. For some hours the confusion of war at its vortex seemed to clear and exhilaration replaced controlled bewilderment.

The battalion's first battle was inevitably drawing closer. Now the stage for it was being set. The bridgehead was expanding in the wake of the thrust by the 17th Division and an armoured brigade. Its way south, down the east bank of the Irrawaddy, was seamed by dry, sandy *chaungs*, which had to be crossed according to tactical doctrine, because each offered the enemy a useful defensive position. Until 4th March, the Japanese had been no more than elusively flitting patrols and stinking corpses. On this day, the company of the 4/8th commanded by Sheil-Small was ordered to make the next move forward.

Walker had accompanied Sheil-Small on his initial reconnaissance and the country ahead seemed clear. Taking the customary precautions, the company moved forward. Then, as Sheil-Small was siting his positions, the blow fell. As Japanese shells and machine-gun fire broke upon his exposed platoons, some three hundred enemy infantrymen burst from close cover in headlong assault. At once the wireless – the link with the rest of the battalion and the artillery – was smashed and violently confused fighting broke out. The Gurkhas had not even had time to dig the shallowest slit trenches.

Walker, out of contact with Sheil-Small, saw the dangers and called for artillery fire. The gunners had to make their calculations from approximate map-references and their shells burst not only immediately in front of, but among and on the embattled company's positions. Several Gurkhas were killed and wounded by the bombard-ment. But more Japanese were hit and they withdrew, the first impact of their assault broken. But they would come again and Sheil-Small set about digging-in and preparing for more attacks, which before long would be in the dark.

The Japanese had had enough. They had been confident that having caught the Gurkha company off-balance and in the open, a combination of a sudden artillery concentration and an infantry rush would carry the position. But the stubbornness of Sheil-Small's stand and the ruthless practicability of the defensive barrage directed around, and sometimes on, the Gurkha positions had driven them back with loss.

Fought in the open, this first battle on the east side of the Irrawaddy was watched not only by Walker but by his brigadier. Seizing on the opportunity to make an example of the dash shown that day, Walker was able to make several immediate awards for gallantry, including one Military Cross to Sheil-Small and another to one of his Gurkha platoon commanders.

Now the 4/8th was involved in regular punching movements for-ward, capturing one fortified village after another. Without the sup-port of tanks and aircraft and closely-controlled artillery fire, the task might have been impossible. Each village had, at its centre, at least one solid, tapering pagoda, constructed so strongly that shellfire only chipped the surface. And around each village was a thick hedge of cactus and beneath this the Japanese would dig their bunkers. While each was being excavated, the cactus would be held back by ropes and, once the defenders and their weapons were inside, these would be cut, so covering the aperture and imprisoning those within.

There was no question of the Japanese surrendering, so that each bunker had to be knocked out, its defenders killed. Sometimes this could only be achieved by driving a tank over it, burying those inside.

Now that the pursuit of the enemy was going according to plan, the headquarters staff of the 14th Army was able to study the behaviour of individual units and this was occasionally irksome to Walker.

Once, on the eve of an attack on a strongly-held village, an Army psychiatrist arrived at battalion headquarters to enquire about morale. Walker suggested that the best way to study this would be to join the attack in the morning. To his credit, the psychiatrist agreed and accompanied Walker, spending some time in a slit trench under point-blank artillery bombardment. There was nothing wrong with morale, he could see, and Walker was delighted that his guest could watch his Gurkhas 'going in like terriers' with drawn kukris.

He was more annoyed by a signal querying reports of enemy casualties. After a battle, only whole Japanese corpses should be counted, ran the signal, so as to avoid over-optimistic body-counts. Walker was told that a senior officer would be visiting him to discuss this.

This signal reached battalion headquarters as a fierce infantry action was ending. As was his custom, Walker washed and changed into a clean uniform, freshly pressed by his batman with a charcoal iron. Then, to welcome his visitor, he ordered that either side of the entrance to the pagoda, which housed his headquarters, should be displayed a severed Japanese head. The visitor arrived, saw the gory heads and said, 'Walker, must you do that?'

'Must you send me silly signals?' replied Walker. 'If I say I have killed a Jap, I have killed a Jap.'

Now came a change of plan. Towards the end of April, the 4/8th were ordered back across the Irrawaddy, to advance south and help cut off the Japanese army retreating from the Arakan.

A formal, set-piece battle now seemed inevitable. Before, the battalion had been involved in mounting attacks and beating off counter-attacks. Now their task would be to cut the enemy off from his base and this would mean standing and fighting with as much tenacity as the Japanese themselves.

The point at which the enemy's retreat was to be cut was the village of Taungdaw. Here the road ran through a valley some fifteen hundred yards wide, dominated by steep hills rising to more than nine hundred feet. Sheil-Small's company was sent ahead to cut the road and hold it, while the rest of the battalion moved up. Knowing strong Japanese forces to be at hand, he decided to concentrate in positions astride the road itself, for he felt that to occupy the commanding high ground as well might spread his company too thinly to face intensive attacks. So Myers was hurried forward to occupy the commanding hill, arriving just as the fury of the Japanese attack fell on both company positions.

So began the three days and nights of the Battle of Taungdaw. A tough Japanese division was determined to break through the road, held by only two companies, and they had been able to seize high ground between Myers and the rest of the battalion. At first, the two isolated companies – dug in and well-stocked with ammunition – threw the Japanese back without undue difficulty. Then ammunition began to run low and casualties to mount. On his hill, Myers had many wounded – some of them with ghastly wounds, stomachs laid open, or limbs blown off – and no means of evacuating them. He was running short of medical supplies, particularly morphine, and as pain-killers were needed for the worst cases, wounded men could only expect a shell-dressing to be laid over their wound and some gaped wide enough to take two.

The only way to relieve the two companies and hold the roadblock was for the rest of the battalion to capture the high ground which the Japanese had seized in Myers' rear. Walker had no alternative but to order an infantry assault. He sent for young Michael Tidswell, one of the most popular of his young Emergency Commissioned Officers, and told him to take the hill regardless of casualties.

An early sign of battle fatigue and nerve failure among officers was, Walker knew, a tendency to question orders, to complain that too much was being asked and to suggest easier means of attaining an objective. But, although Tidswell could see the desperately dangerous nature of his task, there was no questioning of his orders. Like the others, he knew that Walker would only give-such an order if there was no alternative and that he would have been ready to lead the attack himself.

Tidswell led the attack and, as a British officer, was the principal target. Running into strong resistance, his company fought their way up the hillside, taking part of the crest. Tidswell himself was mortally wounded and brought back to battalion headquarters and died on his way to the dressing station. But the Japanese still clung to much of the vital hill.

Walker now ordered another company to attack. This was commanded by a young American, Lieutenant Scott Gilmour, who had been with an American ambulance unit and had transferred to the Indian Army. 'Take that hill, Scottie, and I'll make you a captain on the spot,' said Walker. For the assault he was to have the close support of Hurricane fighter-bombers and, when they appeared overhead, Walker spoke to them by wireless, saying that if the hill was taken he would present the squadron with a Japanese officer's sword.

With bombs bursting only a few dozen yards from the Gurkhas, the assault went in and by dark the Japanese had been swept from the hill. The two isolated companies were relieved, the roadblock held and the Japanese remained cut off.

So ended the Battle of Taungdaw. The battalion had lost ten killed and thirty-three wounded. Two hundred and eighteen Japanese corpses were counted and many more must have littered the jungle, killed by artillery fire and air strikes. A Gurkha, Rifleman Lachhiman Gurung, was awarded the Victoria Cross and there were several immediate awards for gallantry. By any standards, it had been a pitched battle and to the officers of the battalion it seemed astonishing that so many of them had survived. The credit for this must go, they knew, to Walter Walker.

'The Colonel seemed pleased with us,' wrote Patrick Davis, who had fought with Sheil-Small's company, in retrospect. 'He was an extraordinary man, and I use the word extraordinary with care. He was extra ordinary. I knew how I longed for sleep, how my own nerves were worn thin, like an over-used truck tyre, how almost everybody was to a greater or lesser extent, suffering in the same way. I also knew that Walter Walker took less sleep than any of us, could never relax, was for all of every day and much of most nights thinking, planning, encouraging, prodding, looking ahead both in time and space, whether on the move or chained to his command post, never inactive. Yet now ... he seemed much the same as ever. The shadows about his eyes were perhaps deeper, the bones on his face a little more prominent. He was certainly living on his nerves, he must have been, but it hardly showed.'

This was to be Walker's last battle as a battalion commander. Soon after, while working on a map in his command post, Major-General Geoffrey Evans, now the divisional commander, walked in and asked whether Walker thought that his second-in-command, Major Watson-Smythe, would make a good commanding officer. Walker replied that he would. A few weeks later Evans instructed Walker to hand over command and himself come to divisional headquarters as General Staff Officer (Grade One). In vain did Walker and his brigadier protest that this was not the moment to change command, just as the battalion emerged from a period of hard fighting.

It was of no avail. In June, Walker was sent on leave to India – where he managed to spend a few days with Beryl in Quetta – and reported to Evans.

But this was not the end of Walker's association with the 4/8th. A month later, the battalion became involved in heavy fighting for the vital bridge over the Sittang river, which had been the scene of such agony three years before. The ground was flooded, so digging out of the question and the Gurkhas found themselves exposed to accurate shell and machine-gun fire. Heavy casualties were suffered and the conduct of the battle was adversely criticised.

Among those who were acutely unhappy about this action was Watson-Smythe and he sent a private signal to Walker at divisional headquarters complaining that he had not been given the necessary support. The signal came to the notice of General Evans who saw it lying on Walker's desk. When they met Evans said that he had read the signal and that he, Walker, must go down to his old battalion to investigate. The situation he found was not happy. There had been clashes of personality and mutual confidence within the brigade had been damaged. As an emergency remedy, Walker was told to resume command of the 4/8th while at the same time remaining a senior staff officer.

But now the war in Burma was over.

The battle on the Sittang was to be the last of the second campaign in Burma. The Japanese, fighting desperately, were finally outfought and themselves destroyed by British, Indian and Gurkha soldiers as full of zest and vigour as they themselves had been three years earlier. On 5th May, 1945, just as the war in Europe was ending, 15 Corps of the 14th Army entered the abandoned capital, Rangoon.

By the beginning of August, the Japanese had been cleared from the west bank of the Sittang and the 14th Army began regrouping for the next phase. The war in Europe having ended three months earlier, ample shipping, landing craft and amphibious forces were available and their use could avoid the head-on collisions of land opponents. A major sea-borne assault on Malaya was to be launched across the Bay of Bengal to cut off both the Japanese armies in Siam and in Singapore. But before this could be launched, the war – and future wars – was transformed. On 6th August, the atomic bomb was dropped on Hiroshima.

A second atomic bomb was dropped over Nagasaki on 9th August and, next day, Japan accepted defeat.

The 14th Army had inflicted upon the Japanese the greatest and most decisive defeat they were to suffer during the Second World War. But this was not the time for resting on laurels – for Walker, the Distinguished Service Order – there were the immense areas of territory captured by the Japanese early in the war to be re-occupied and the rescue of many thousands of Allied prisoners in prison camps through the East.

The task of the 7th Division was to be the occupation of Siam.

It was ironical that these first weeks of peace were for Walker – and many others who had survived so much – to be the time of most acute danger. For some it would be the unexpected struggle with nationalist or communist guerrillas, who themselves hoped to inherit the Japanese conquests and were determined to thwart their former colonial masters. For others it was to be the routine risks of the turbulence stirred up by the collapse of armies. For Walter Walker it was to be the simple matter of flying from Rangoon to Bangkok.

Divisional headquarters was to follow General Evans to Bangkok. Walker had a Dakota transport aircraft allocated to him, his staff, equipment and records. Take-off was, as usual, early in the morning before the dangerously turbulent cumulus cloud had built up under the heat of the sun. At first, the flight was uneventful. Then half-way between the two capitals, the pilot came back to tell Walker that one engine was racing and that he might have to attempt a crash landing. The crew had parachutes, the passengers had not, so Walker and his companions busied themselves stacking their kit and baggage near the door, ready to jettison should the danger increase.

All was well. The Dakota landed safely back at Rangoon and its load transferred to another aircraft. It was now late in the day and the gleaming white, black-hearted thunderheads had grown dangerously.

At a time like this, there could be no thought of postponing so important a flight because of suspected bad weather and again Walker took off for Bangkok.

The Dakota circled to its maximum height of about twenty thousand feet but still the cloud towered about it and there was no alternative but to fly straight towards Bangkok. The course of the aircraft took it through the heart of an electric storm. Never before – not even in the fiercest battles with the Japanese – had Walker been so certain that death was imminent.

The Dakota was flung about the sky in blinding flashes of blue lightning. It seemed impossible that it could survive. Others, indeed, did not. Another Dakota, loaded with British infantry, vanished in this storm, as did another, flying in the opposite direction, tragically, carrying one of the first parties of prisoners of war to be released in Siam.

Eventually the landing was made at Bangkok, where an impatient General Evans was waiting to ask why on earth his staff was so slow in joining him. In Siam, the end of the war had not brought complete peace. The Japanese there had not been defeated in battle and showed their pride and arrogance. On occasion, their convoys would try to force British vehicles off the road. While Walker was in charge of negotiations for the formal surrender, the 4/8th Gurkhas, also in Siam, took on the depressing task of guarding prisoners from the Indian National Army, Indian soldiers who, after capture in 1942, had thrown in their lot with the Japanese. At this time, Peter Wickham, the adjutant of the 4/8th, saw Walker's brother – one of the released prisoners – and could report that he was still just alive.

The battalion was now sent to Malaya, then to Java to take part in the repression of the nationalist guerrillas and finally back again to Malaya. Now Walker returned again to take command, with Peter Myers, who had decided to stay in the Army, as his second-in-command.

In 1946, it seemed that the war had been over for a long time and the sunlit prospect of peace seemed to lie ahead as far as the imagination could see. In many regiments, officers settled back gratefully into the pleasant routines and social and sporting rounds of peace. In the 4/8th Gurkhas the commanding officer worked himself as hard and expected as much of his officers as ever he had in war. Night after night the light on his desk would burn until three in the morning.

A perfectionist, he was preparing for peace as intensely as he had prepared for war. And, with the collapse of the British, Dutch and French eastern empires imminent, he was also preparing for other wars beyond the peace.

4

Emergency in Malaya

The twelve years of guerrilla war that were known as the Malayan Emergency had an official beginning and end but the reality was not, of course, so neat. The gradual slide to conflict had begun a long time before, perhaps as long as a quarter of a century, when the Chinese settlers in Malaya imported their politics as well as their industry. So while the Kuomintang Nationalist Party was predominant there, as in the homeland, the Communist Party also grew among the immigrants, as it began to spread roots under the care of Mao Tse-Tung. Significantly, as well as nationalism and communism, the Chinese brought with them their own unique and dangerous form of political and criminal intrigue: the secret societies.

From its inception, the Malayan Communist Party was partly a popular movement and partly a secret society. It, too, had links with the Comintern and the Soviet Union and it was on instructions from Moscow that, in 1939, the Chinese Communists in Malaya did not support the British Empire in its stand against Germany but, like Communists everywhere, changed their mind when Hitler attacked Russia. Japanese atrocities in China made them the natural enemy of the Malayan Chinese and the Communists were also influenced both by Russian antagonism towards Japan and by the weird fascism of emperor-worship. So it was that the British, looking back at the Malaya from which they had been driven, saw the Communists there as potential allies. And so they became.

It began in a small way. A few British officers were filtered back into Malaya to try to establish a resistance movement. This gradually evolved as the British Force 136 and as the Chinese Malayan People's Anti-Japanese Army.

Had the Americans not exploded the atomic bombs over Japan in August, 1945, it is possible that the well-armed Chinese guerrillas in the Malayan jungle would have risen against their oppressors and significantly helped the British sea-borne invasion. As it was, they were able to claim the victory without having had to fight for it.

As British rule was gradually reimposed, the guerrillas were officially demobilised with gratuities and, it was hoped, disarmed. But they were not fully disarmed, any more than they forgot their military training or abandoned the military formations in which they had been living. So it was that, three years after the end of the Second World War, there was still a guerrilla force, armed and organised, awaiting the call to action.

This came at the beginning of 1948. In February of that year, the Comintern held a conference at Calcutta and, it is now thought, that it was then decided that the time to strike in South-East Asia had come. The European and American colonialists had not yet had time to re-establish themselves in their old empires and their prestige in the eyes of Asians had not been restored. In the colonies themselves, both nationalist and communist movements still commanded popularity and retained some of the war-time spirit and organisation that would be necessary for revolutionary war.

Burma, on the verge of independence from Britain, seemed the easiest prize and proved to be so. Indonesia could never again be the Dutch East Indies and was so large and populous that perhaps only the discipline of communism could provide a stable framework for the exploitation of its resources. In Indo-China, the communist Vietminh was already locked in battle with the French colonialists. But the keystone of South-East Asia was Malaya and its great dependent port and commercial centre of Singapore.

The fifty thousand square miles of Malaya were rich in tin, rubber and timber but this was not the prime reason for its importance in the eyes of the Comintern. Malaya was one of the brightest jewels in the imperial crown, well-ordered and run with British flair and justice. The most stable of all the European colonies, its fall would soon be followed by the rest of the prizes.

Of Malaya's Asian population of more than five million nearly half were Chinese settlers: industrious, intelligent and tied emotionally and economically to China itself. Their loyalties, it was thought at Calcutta, would be towards their homeland, their race and their families and certainly not towards a monarch in London. The remainder were Moslem Malays, the indigent race: charming but lacking the ambition and thrust of the Chinese. These had been, and were still, ruled by some twelve thousand Britons – civil servants, policemen, businessmen, planters and mine managers – whose power was on the wane.

Early in 1948, Malaya was granted a first measure of independence as the Federation of Malaya. Nominally self-ruling, it had a British High Commissioner and it was he who dictated policy in foreign relations, defence and internal security. Meanwhile Singapore remained tightly under British rule as a Crown Colony.

Malaya had been relatively content under British rule and the successful Chinese settlers, who as traders dominated the economy, could not be expected to rebel. But there was a large class of discontented Chinese, who could be expected to listen to talk of rebellion and to act. These were the six hundred thousand squatters – landless Chinese, living off the edge of the jungle, many of them having fled there from Japanese persecution and the haphazard miseries of war. They were the power base upon which the revolution could be built.

While military preparations for revolution were hastened there was much else that could be done. As a matter of routine, Chinese communists had worked and intrigued their way into key positions in the trade unions and, first, it would be their turn to act. Throughout the first half of 1948, Malaya was racked by a series of violent labour upheavals. Disputes over pay and working conditions were worked up into violent confrontations, often leading to rioting and sometimes death. Industry was disrupted across the country.

What particularly worried the employers was that the violence and intimidation came not only from militant trade unionists and workers but from outsiders who slipped into the rubber plantations and tin mines from the jungle and vanished before the police could be summoned to intervene. Soon these were being described as 'communist terrorists' or 'CTs' and it was they who increasingly emerged as the enemy that would have to be fought.

The British, who had personal stakes in Malaya, called for strong action from the High Commissioner, Sir Edward Gent, who was responsible for the internal security of the Federation. At least, they demanded, he could begin by banning the Communist Party. But Gent, on instructions from the Labour Government in London, was determined to play as cool a hand as possible and demurred. Violence increased and, with it, the frustration and anger of the mine-managers and planters.

Finally, as was inevitable, came the moment when open conflict could no longer be postponed. On June 16th, Chinese communist terrorists visited the plantations around Sungei Siput in the northern state of Perak. There, in cold blood, they shot and killed three English planters. That day, Gent declared a State of Emergency in Perak, to be followed next day by its extension to the whole of Malaya and, the following month, to Singapore. The war had begun.

When the crisis tipped over the edge from turbulence to turmoil, Walter Walker and Beryl were on leave. The climate of Malaya changes little with the seasons but by taking to the mountains, which rise to seven thousand feet in the central range, there was a chance of fresh breezes and cool nights. The Walkers had chosen one of the nearest mountain resorts to Kuala Lumpur, in the Cameron Highlands, the nearest to an Indian hill station that Malaya could offer.

There, on the evening of 16th June, Walker received a telephone call. It was an urgent recall to duty.

Since the end of the war against Japan, Walker's career had been one of quiet but steady success. He had presided over the dispersal of a division in India and then, in 1946, there had been his first home leave for ten years. They had returned to visit their families in England with the twins, now bounding with energy and speaking little but fluent Gurkhali. This was not the England the couple remembered from the 'Thirties: now it was a shabby, restricted, edgy place far removed from the comforts still taken for granted by the British in the East.

Now followed an appointment back in India. Walker was chosen to succeed Lieutenant-Colonel Jack Masters – later to become famous as John Masters, the writer – as the Senior Staff Officer (G1) to the Director of Military Operations at General Headquarters in Delhi. Here he served both Lord Wavell, the Viceroy, and Field-Marshal Sir Claude Auchinleck, the Commander-in-Chief, and on one occasion was to write an urgent strategic appreciation based – without their knowledge – on those written by them and Slim when they had been Directors of Military Operations.

While India was convulsed by religious partition and Moslem fought with Hindu, the General Staff was concerned with transferring responsibilities from British to Indian officers. The intelligent young Indian lieutenant-colonel who took over from Walker was Sam Manekshaw, later the Indian Chief of the Army Staff during the war with Pakistan in 1971.

The future of the Gurkhas was under discussion and their British officers were being asked to opt for the British regiment they would prefer to join, should none be left serving with the British Army. Walker chose his old friends the Sherwood Foresters and, for a time, wore their regimental badge.

When GHQ was finally transferred to Indian control, Walker's next appointment was made known to him. He was to be G1 in Malaya District Headquarters in Kuala Lumpur. After some months, he and his family took leave in the Cameron Highlands and it was there that he took the telephone call that called him to arms. An armoured car would escort him down to the Kuala Lumpur road, he was told. The caller added, 'The whole country is about to go up in flames.'

In retrospect, it seems extraordinary that more practical preparations for the coming storm had not been made. On paper, the garrison of Malaya was reasonably strong. Major-General Charles Boucher, the GOC, Malaya District, had six Gurkha battalions on the mainland – strung out among the rubber-planting and tin-mining regions in the west – with another on call in Singapore and an eighth away in Hong Kong. One British field regiment of the Royal Artillery was also on the mainland training Gurkhas as gunners and there were three British infantry battalions, two on Singapore island and one at Penang.

The Gurkhas themselves, while potentially as formidable as they had always been, were not at their customary peak. Most of those who had fought in the Second World War had returned to Nepal and in the four of the ten regiments, which had not been transferred to the Indian Army at the beginning of 1948, the process of re-forming and retraining was still in progress.

Soldiers were only to be used in support of police action and, at this time, the Malayan Police Force was at a low ebb. Its establishment was ten thousand – mostly Malays – but it was in no way prepared to face insurgency. There was no Special Branch and any counter-subversion duties fell to the Criminal Investigation Department. Thus, what were to become known by the collective title of 'Security Forces' lacked hardened muscles and alert eyes. A symptom of the general attitude was that, particularly at this early stage, the insurgents were more often known as 'bandits' than as 'communist terrorists' or 'CTs'.

But, within a month of the declaration of the Emergency, a practical step was taken; one that showed that the dash and originality, that had been at so high a premium only three years ago but was being forgotten so quickly, was not yet beyond recall. Remembering the lessons of the various Special Forces, which across the world, had learned to take the initiative with ferocious subtlety, often beating the enemy with his own tactics, Ferret Force was raised. This, as its name implied, was designed to seek out and expose the guerrilla enemy so that he could be brought to battle by the main forces.

Former British officers of Force 136, many of them now planters and so eager for action, were asked to volunteer. Ironically, they would now be leading parties of British and Gurkha soldiers, Malay and Chinese civilians and even Dyak trackers, brought over from Sarawak, against many of the Chinese guerrillas they had trained and led against the Japanese. Ferret Force was to be formed at the Malayan Regiment's depot at Port Dixon and its Commanding Officer was to be John Davis, who had been a close friend of the rebel leader Chin Peng in the jungle, when they had been together in Force 136. To train and equip this swashbuckling force of individualistic irregulars, a smart and professional soldier was chosen: Lieutenant-Colonel Walter Walker.

Not for the first or last time came the confrontation between professional and amateur. Some of the hastily-commissioned civilians of Ferret Force had more self-confidence than their experience in Force 136 and their experience of the Malayan jungle justified. Their adventures – real or anticipated – of 1944 and 1945, sometimes made it hard to realise that their immaculate training officer had had far more experience of jungle operations than any of them. For much of their own war in Malaya their primary objective had been their own survival and their offensive actions had been against the installations and communications of Japanese armies engaged in a campaign hundreds of miles away from the scene of their own endeavours.

Walker's own study of communist terrorist strategy and tactics convinced him that he would be dealing with an increasingly powerful and cunning enemy and that nothing less than the methods evolved to fight the Japanese themselves would be effective. From General Boucher downwards, he noticed a confidence and over-optimism that he thought dangerous. Soon after taking up his new duties at Port Dixon, he had a chance to demonstrate this.

One of the Ferret Force officers – a tin-miner who had been with Force 136 – was outspoken in his criticism of Walker's tactics, which he regarded as far too intricately planned and rehearsed: sledge-hammers to crack nuts. In particular, he considered Walker's ambush procedures to be unnecessarily elaborate, himself tending towards more impromptu methods reminiscent of the Wild West. He had to be taught a lesson and Walker asked him to lead a patrol of nine men from the training camp into the jungle, using the methods he had advocated. They had hardly moved one hundred yards down the jungle trail when Walker sprung his own ambush and, had the ammunition been live, not one of the patrol would have survived.

A typical Ferret Force group, led by a British volunteer with local knowledge, would consist of four teams, each consisting of a British officer, twelve volunteers from British or Gurkha regiments, the Malay Regiment, a detachment of the Royal Signals, Dyak trackers and a Chinese liaison officer. The teams travelled as light as they could and only essential rations were parachuted to them as they tried, as much as possible, to live off the land – jungle specialities sometimes including fried iguana lizard, python soup and roast tortoise.

Walker had been putting these teams through their rigorous training for only about a month when the depot was visited by General Sir Neil Ritchie, who had been an unsuccessful commander of the 8th Army in the Western Desert, a successful corps commander in North-West Europe and was now Commander-in-Chief, Far East Land Forces. Walker arranged a demonstration of an attack on a terrorist camp – including the silent killing of its sentries – for the occasion and when it was over, Ritchie led him aside. It had been an admirable display, he said, but Walker was going to be moved to another, more important job.

Major reinforcements were being sent out from Britain. They were some of the best infantry in the Army but they totally lacked jungle experience. Indeed, according to tradition, they had never served East of Suez. The reinforcements were to be the Guards Brigade. To acclimatise and train, they were, on arrival, to go straight to the Far East Training Centre at Johore Bahru. This was still largely a conventional training centre and a jungle warfare wing had only recently been opened. Walker was to take over as Commandant and gear the whole establishment to the preparation of British soldiers to fight jungle guerrillas on their own ground.

Ferret Force survived only a matter of weeks after Walker's departure. So mixed a force inevitably had its political difficulties. There were disagreements between Army and Police officers over its methods and its use and regular infantry battalions were often reluctant to lose some of their best individuals to this 'private army' just as they themselves were about to be put to the test. So, before the end of the year it was disbanded. But its brief life was not wasted. It helped inspire not only the deep-penetration operations of the Special Air Service at a later stage of the campaign but also the routine jungle patrolling by infantry platoons, which would otherwise have seemed hopelessly beyond the capabilities of European soldiers.

Walker's first task was to convert guardsmen into jungle fighters. The Jungle Warfare Wing of the Far East Training Centre was housed in a former mental hospital at Tampoi, chosen because it was within easy reach of a variety of jungle terrain. Before his pupils arrived, Walker himself covered the ground by day and sat up late writing scenarios for exercises, covering as many likely jungle fighting situations as possible. Finally, there was his demonstration platoon to be trained.

The first arrivals were the brigadier – soon after to die in a light aircraft accident – and his battalion and company commanders. Walker planned to put officers and senior non-commissioned officers through the jungle training course before their men so that they themselves could carry out most of the training of their battalions.

When the Guards had moved on, regular six-week courses continued. After Walker's opening address, his pupils would attend a lecture followed by a demonstration and then try out the technique themselves.

The climax of each course came about five days before it ended, when every class under instruction was sent into jungle where communist guerrillas were known to be operating. One class found itself face to face with three armed terrorists and, to its own surprise, killed them all.

The confidence which Walker felt in the trained soldiers he returned to the fighting battalions was not extended to most of those in charge of preparations for war. He had a chance to make his views known in a private talk with Malcolm MacDonald, the son of Ramsay MacDonald, who was the British Commissioner-General for South-East Asia. MacDonald told Walker that British generals in Malaya seemed confident that the crisis was under control and that the arrival of the Guards would bring about a quick victory. Did Walker agree? Walker replied that he did not. For a start, the British Army was geared to fight a nuclear, or conventional, war in Europe and all the thinking at the Staff College and the School of Infantry was in that direction. Moreover, it seemed to Walker that all the skills learned and equipment developed in the long struggle with the Japanese had been lost. The Army would have to learn all over again.

It was agreed between MacDonald and Walker that something might be achieved if a demonstration of jungle fighting were arranged for himself and senior officers and for Walker to deliver a commentary loaded with criticism. This duly took place and reaction was immediate.

One demonstration was of the ambush of troop-carrying lorries and spectators watched a convoy of 'soft-skinned' vehicles trapped and destroyed by guerrillas. This was followed by a demonstration of ways of avoiding such disaster and, in his commentary, Walker pointed out that troops were being asked to travel long distances in unarmoured trucks and were not equipped with the portable flame-throwers or grenade-dischargers for rifles: both being weapons used with success against the Japanese.

After the demonstration, General Boucher and Colonel Nicol Gray, the Commissioner of Police, complained to Walker. Putting troops in armoured vehicles was, said Boucher, too 'defensive minded', Gray agreeing and adding that if he thought Walker meant what he was teaching he would send no more police officers to the Jungle Warfare School. In vain, Walker explained that these were the tactics that had been learned so painfully in Burma but seemed to have been forgotten since 1945. But chance and the Communists themselves were about to make Walker's point for him. That night, while Boucher and Gray were returning to Kuala Lumpur a police party in a 'soft-skinned' vehicle was ambushed on the road and wiped out.

The making of public complaints was clearly a successful trick and Walker tried it again, this time in the presence of General Sir John Harding, then Commander-in-Chief, Far East Land Forces. Walker had got his flame-throwers but the cartridges used to ignite the Latex fuel were unreliable and labelled, 'For training only'. These were the weapons upon which soldiers would have to rely to survive a fierce, close-range ambush. Before the demonstration began, Walker emphasised the value of flame-throwers, adding that unfortunately he could not guarantee that the Latex would ignite. The demonstration began, the ambush was sprung and not one flame-thrower threw flame.

On the spot, Harding asked a senior staff officer to ensure with urgency that reliable cartridges were made available.

Harding understood the purpose behind such making of direct complaints and he was, perhaps, watching for any sign that this forceful and cunning Commandant had over-reached himself. Later, he thought he had found one. Walker's drill-sergeant from the Guards had had all the tent-pegs of the visitors' tent white-washed and that seemed just the sort of useless, time-wasting punctilio that would be picked upon and lampooned by the popular newspapers at home. 'Why,' demanded Harding, 'are you wasting time white-washing tent-pegs?' 'Because,' replied Walker blandly, 'we do this demonstration at night and I do not want the spectators to trip over them.'

Within a year, Walker had laid the foundations of what was to become world-famous as the Jungle Warfare School and eventually, twenty-two years later, be transferred to Malaysian command. Its value was shown not only in increased success against the communist terrorists in the jungle but in the attitude of the British soldiers in Malaya. The jungle was losing its terrors for the young conscripts fresh from home and learning to live like a beast of prey became an experience both exciting and routine.

After thirteen months, Walker handed over command in November, 1949, and returned to England for a six-month course at the Joint Services Staff College. At Chalfont Latimer the studies concentrated upon Europe and the threat of Russia. Berlin was under blockade and, in response to the outbreak of what was to be called the Cold War, the North Atlantic Treaty had been signed and the NATO military alliance founded. But Walker found his thoughts constantly strayed from Europe to Malaya. After leaving Tampoi he had been awarded the Order of the British Empire but he was more eager to practise what he had preached at the Jungle Warfare School than to debate the theoretical strategic implications of nuclear warfare in Western Europe. In Malaya there were practicalities awaiting his attention.

His impatience was soon rewarded and in the most stimulating way he could have wished. He was told that, on his return to Malaya, he would take command of a battalion, the 1st Battalion of the 6th Gurkha Rifles.

Walker was back in Malaya by the autumn of 1950 and on his way to Bahau in South Negri Sembilan where the 1/6th were deployed in rubber plantations around the railway line from Singapore to central Malaya. Paying the customary calls on senior officers of the hierarchy above his own command, Walker heard of the reputation that he was about to inherit and he was both concerned and stimulated by what he was told.

In most ways his new battalion was good. Certainly it was extremely smart. But on operations it had not been a success. For a battalion of its undoubted quality, the 1/6th had been a disappointment. Obviously, Walker's first task would be to find out why.

When he reached his battalion headquarters, he saw that the companies were all out in the rubber estates and that the only way to assess them was to go out and live with each in turn. This he did, for ten days at a time, and, as he expected, he found his answer. The 1/6th were tired. Their ambush parties went out but they failed to kill their enemy and, to Walker's eyes, the reasons were clear. Fatigue had induced carelessness. They took too many risks; they skipped essential precautions; they were noisy. But the material of the battalion was good and it was Walker's task to make it as efficient as it deserved to be.

In October, after fifteen months of operations, the battalion was ordered north to Sungei Patani in Kedah for rest and preparation for the next spell in the jungle. Walker now had two months in which to mould the battalion as he wished. They were very tired, so should they rest – at least for a time? Or, far from allowing them to relax, should he drive them even harder in their training than he had on operations? Walker chose the latter course and, as he said, he 'ran roughshod over the battalion' and he 'hammered them'. There was no let up and officers who had looked forward to languid days on the beach at Penang found themselves back in the jungle practising again and again the skills of patrolling and ambush that they thought they knew so well and doing it under the ruthless eye of a perfectionist.

Before the two months were up, Walker was driving the battalion so hard that word was discreetly passed to him that his officers were on the verge of mutiny. He disregarded the warning, having seen for himself the signs of returning self-confidence.

Not all the officers of the 1/6th were happy with their new Commanding Officer. They had settled into a routine of operations based on the most obviously convenient means of making contact with the enemy. Cordons would be set up along the line of a road so that the troops could be transported there more easily. Artillery fire would be put down on jungle areas without any identified target in an attempt to deter the terrorist from occupying it. Now Walker was not only demanding new and more difficult tactics but insisting that the officers took part in evolving them.

During their demanding indoctrination the battalion was visited by Major-General Roy Urquhart, General Officer Commanding Malaya District. He had commanded the Sixth Airborne Division at Arnhem and his tough but not always imaginative approach to soldiering had earned him the nickname of 'The Sheep-Stealer'. Urquhart heard the complaints that Walker was driving his battalion too hard and he seemed to believe them.

Some time later Walker received his Confidential Report for 1951 and was appalled by what he read. This not only implied that his methods were too harsh but that his battalion was now not as good as it had been. This damning report arrived just as the battalion was demonstrating the value of its recent hard training. Walker at once wrote to his divisional commander saying that he refused to accept this verdict and his formal representation was passed on to Urquhart. Six months later Urquhart again visited the battalion and next day wrote to Walker saying that he would initiate a fresh Confidential Report, 'for any doubts which I had six months ago have now been removed'.

By the end of the year, the 1/6th were back on operations, their fatigue converted to resilience, their skills restored and sharpened. Their new territory was in the low-lying swamps of Johore in southern Malaya. It was clear to Walker that they were now well-attuned to the jungle, but he decided to seek a second opinion. Instead of himself accompanying the companies, as he had on arrival, he sent with them a surrendered Communist guerrilla with orders to report back to him, fully and frankly, on the efficiency of each. As a result new tactics were evolved. The guerrilla taught the Gurkhas how to hide their tracks, not only by the last man in a patrol sweeping them away with leaves, but by such tricks as crossing a stream backwards and purposely leaving the footprints.

Malaya during the Emergency

Walker was now summoned to Kuala Lumpur by General Sir Hugh Stockwell to help with the preparation of a new military training manual, *The Conduct of Anti-Terrorist Operations*, as the man who, more than any other, had taught the British Army in Malaya how to fight.

In April 1952, the 1/6th were ordered to move again, this time to the hills of northern Perak between Ipoh and the border of Thailand, where they were to relieve 40 Commando, Royal Marines. This was dangerous country for it had been at Kuala Kangsar that the Communist rebellion had begun. The terrorists were not only well-established in camps deep in the jungle-covered hills, but dominated the roads.

Walker's first move was to examine his predecessor's tactics. The Marines, whom he always admired, seemed to have worked as heavy infantry and kept to a formal routine. Walker disagreed with most of this and determined to reverse their methods. While the Marines tended to relax at weekends, the 1/6th would then put in their maximum effort. The Marines had tried to control the jungle from the road so Walker decided to try to control the road from the jungle.

The 1/6th had its headquarters at Kuala Kangsar, where the Sultan of Perak had his Istana, and detachments out at Grik, Lenggong and Taiping. One of Walker's first moves on arrival was to establish close liaison with the Police – particularly with the Special Branch – because it was from them that much of the intelligence information, upon which his operations would be based, would have to come.

The head of the Special Branch was Cyril Keel, who had a reputation for a sure and subtle touch in finding the weak links in the Communist system, finding potential informers and manipulating them. Keel and Walker now began a successful partnership.

One evening at Kuala Kangsar, Keel invited Walker to his house and there introduced him to a Chinese of insignificant appearance. His name was Leong Cheong and he was the District Committee Member for the terrorists in Lenggong. Keel had discovered, through his family, that he might be willing to betray his organisation for money and so it had proved. Code-named 'Jonquil', he was to be paid a substantial retaining fee and rewards for information that led to a killing.

Information began to flow at once. Early in June a surrendered terrorist was identified as Wong Siew Lam, a doctor with the 12th Regiment of the Malayan People's Liberation Army, and he offered to lead Keel to a hospital camp in the jungle where the regimental commander himself might be found.

An operation was mounted. The surrendered doctor led a company of the 1/6th through the jungle to the approaches of the hospital. One platoon stealthily moved round the flank of the enemy camp to act as cordon while the assault platoon moved up ready to charge from close range. Walker, who was with the latter, soon saw the tops of huts through the trees and heard Chinese talking and singing.

The 'stop parties' were taking up ambush positions to fire upon fugitives from the camp and the assault section had crept to within seventy yards of the camp when there was a burst of firing. Several short bursts of automatic fire and three single shots and surprise was lost. The singing stopped instantly and when the assault section charged into the camp it was deserted. The Chinese guide then told Walker that the firing of three single shots had been a known guerrilla signal meaning 'a comrade has surrendered' and that the bursts of automatic fire probably meant 'evacuate the camp immediately'.

Although this attack failed, others succeeded. Of these the most remarkable was the direct result of a dangerous mission into the jungle undertaken by Walker and Keel.

The Communist informer 'Jonquil' had sent word that he was ready to lead the security forces to an important objective but that he must personally give the preliminary details to senior officers at a rendezvous in the jungle. Walker was asked if he would be willing to accompany Keel on this mission and accept the risk that it might be a baited trap. Walker agreed and it was arranged that on a day when a particular road was generally closed to traffic they would take a taxi down it and then head out in the jungle.

They set out, wearing civilian clothes, and taking the precaution to have with them a Gurkha signaller, also in civilian dress – but carrying a wireless set in his pack and a Chinese interpreter. If they were to be ambushed, at least their last act could be to call in the Gurkhas on to their assailants.

At the prearranged point on the road the taxi was stopped and the four men set out into the undergrowth. When they were nearly a thousand yards from the road, they sighted a small hut beside a stream, which, Keel whispered, was the rendezvous.

Leaving the signaller hidden but with a view of the hut, the two men, followed by the interpreter, crept forward, pistol in hand. The hut was empty and Walker felt a sudden dread that this was to be an ambush and that a burst of machine-gun fire could hit them at any moment.

But it was not an ambush. After about five minutes, there was a rustling and Leong Cheong was beside them, grinning. The whispered information was quickly interpreted. A group of six important Communist guerrilla leaders were about to pass through the area and would stop briefly at a certain abandoned terrorist camp deep in the jungle. But Leong Cheong would soon know more, so another meeting must be arranged.

As Walker and Keel approached their second rendezvous with Leong Cheong, they felt even more apprehensive than they had before. The jungle had an uneasy feeling and, to minimise the effect of an ambush, the two men hid on separate hillocks within sight of one another. There was no sign of 'Jonquil'. Then, Walker's battalion headquarters came through on the wireless with the curt message, 'You are surrounded. Stay where you are. We are sending a company in to rescue you.'

It was comparatively open jungle and Walker strained his eyes to see whether the first movement he would see among the trees would be of terrorists or Gurkhas. It was Gurkhas, moving as stealthily and silently as he had trained them. As they came within earshot, Walker, keeping in hiding, gave a bird-call signal. The Gurkhas froze in their tracks. Walker repeated the signal and his men moved forward. Meeting the company commander, he was told that a message had come through from Leong Cheong to say that he feared he was under suspicion and that the plan for a meeting must be abandoned until he again felt secure. Therefore, Walker's second-in-command, fearing that the informer might have confessed and that the jungle rendezvous might be known to terrorists, decided to act quickly.

Later another message from the informer came through and this time it was what Walker had been hoping for. It was the location of the deserted jungle camp and the approximate date – period of four days – during which the terrorist leaders could be expected to arrive there. Walker at once sent out a patrol to find the camp site and survey it so thoroughly that, on their return, they could make a sand-table model of it. They were also to reconnoitre the various approach routes and survey them so that Walker could prepare the trap he hoped to spring.

The patrol was successful and, on its return, Walker was optimistic but planned to leave nothing to chance. With his bitter memory of the failure of the earlier attack he planned to use seven platoons to make certain of killing the six enemies he expected to find.

Security was tightened as much as possible and an elaborate feint arranged. After concentrating at Lenggong, the battalion moved by night to the Perak river. This was crossed during darkness in boats with muffled oars manned by police. On the far bank, they began a two-day march through the jungle to the camp, arriving there on the evening of the second day, the eve of the first possible date of their quarry's arrival.

Cautiously Walker and the commander of his original patrol entered the camp, wearing light shoes and brushing away their footprints as soon as they made them. Its lay-out was exactly as it had been described. Walker's problem was to entice the six terrorists out of the dense cover surrounding the camp and into a killing ground. They would, he thought, probably make for a hut in the small clearing and here was the clearest field of fire. But in order to produce this concentration of fire, the killing ground must be covered by as many Gurkhas as possible yet the enemy would have to walk through their cordon, unsuspecting.

The six platoons he planned to use as a cordon round the camp would have to be heavily concealed. The assault platoon, which Walker himself planned to accompany, would be hidden very close to the camp but the problem would be to remain concealed and yet know when the enemy had reached the hut. Then Walker noticed a small patch of scrub in the middle of the clearing near the hut. This, plus superb fieldcraft and some cool courage, could provide the answer.

Walker asked for the best section commander from the assault platoon and one other Gurkha. These he ordered to hide in the scrub and be prepared to stay there for four days without making a sound or moving, even to relieve themselves. They would have a light machine-gun trained on the hut and a thong of vine tied to a Gurkha's wrist and stretching across the clearing to the wrist of the officer commanding the assault platoon. When the two Gurkhas first saw the enemy enter the camp they were to tug the vine to alert the attackers. Then, when all six of the terrorists had entered the killing ground by the hut, they were to open fire on them with the machine-gun and immediately the assault would go in.

By dawn next morning all seven platoons and the two hidden Gurkhas were in position. Walker settled down for a long wait. Then, after four hours, he heard a distant shot and, soon after, monkeys began chattering in the trees. Somebody else was about in the jungle.

For the men in the listening-post this was a time of extreme tension. They lay beside their machine-gun, facing the hut. They knew that the most likely path the terrorists would take into the camp was behind them but, as they dare not look round, they could only hope that their camouflage would save them from being seen and shot in the back.

The Gurkhas heard a faint rustling of leaves and a Chinese padded softly past them, looking warily about, his finger on the trigger of his automatic weapon. Another followed. And another. They reached the hut in silence, again searching the jungle with their eyes. Then one, who appeared to be their leader, nodded and they began to shed their packs. The leader now began to speak, as if telling his companions what their defensive positions would be. Suddenly, they froze, silent and motionless. The leader's eyes were fixed on the clump of undergrowth in the middle of the clearing. Slowly, gun in hand, he walked towards it.

Walker, lying hidden among his Gurkhas, strained his ears. A burst of automatic fire hammered in the clearing and he was on his feet, the platoon commander shouting, 'Charge!'

The battle ended in a matter of seconds. The machine-gun in the listening post killed two of the terrorists, the others turned and ran into the jungle and into the sights of the cordon. They were instantly cut down. The entire party, including a senior committee member in Perak and a political commissar, had been killed. The operation had been as perfect as it had been daring. Walker made his report by wireless and this was forwarded to Kuala Lumpur and Singapore and also to General Templer, who was in London.

This success was repeated in August. Then a platoon of the 1/6th under a Gurkha lieutenant, acting on information received from the informer by Special Branch, laid an ambush on a jungle track. Six terrorists walked into the killing ground and six died there.

Operation Hunter – as this sequence of operations was known – continued into the beginning of 1954. But then it lost its fount of information, 'Jonquil'. Leong Cheong had become careless. After his earlier fears that he was suspected by his comrades, he again felt secure, continued to keep secret rendezvous with Keel and to receive his blood-money.

The loss of at least twenty guerrillas – many of them senior committee members – had, of course, resulted in the gradual assembly of a short list of suspects and Leong Cheong was among them. Finally the time for questioning him arrived. First, he had to be searched, more as a matter of security routine rather than in any expectation of finding incriminating evidence. But this there was. Not only had he unexplained money on him, but Leong Cheong was carrying a note recording his last payment from Keel.

The death penalty was inevitable. But his treachery had been such that it was decided he should be made a dreadful example. In the presence of his comrades, Leong Cheong was tortured to death and he was a full day dying. His torturers did not realise that their ghastly work was, in fact, to be counter-productive. Two of the young guerrillas, who were required to watch, were so repelled that they decided to surrender. When, soon after, they did so, the Special Branch 'turned' them just as they had Leong Cheong, and they became the latest in the long succession of informers who helped defeat their own revolution in Malaya.

Operation Hunter had ended in triumph for Walker and now he was ordered to move his battalion from Kuala Kangsar to Ipoh to take over an area including Sungei Siput, where the killings that had marked the official beginning of the Emergency had taken place. Fresh from his successes, he was disappointed in the quality of the civil authorities with whom he was to work. The State War Executive Committee lacked enthusiasm and the police appeared sluggish by his standards.

Now General Sir Gerald Templer, the High Commissioner, visited the battalion at Sungei Siput. Walker had met him once before at Kuala Kangsar and had been impressed by his intelligence and zest. At once, Templer wanted to know what progress was being made in the area. Comparatively little, Walker had to confess. Why? Walker had a lot he could say but, at first, held back. Sensing this, Templer pressed for a full and frank answer and Walker gave it. He complained of the lack of initiative and the poor quality of some of the civil authorities. One local policeman, he said, was nicknamed 'Kipper' because he was two-faced, had no backbone and he stank.

Templer told Walker to travel with him in his car on the drive back to Ipoh and to continue with his complaints and suggestions. On arrival he announced that he wanted to see the State War Executive Committee immediately. The brigade major thought they would all be playing golf but, noting the tone of his voice, collected them for a tongue-lashing from the High Commissioner. This had its usual galvanising effect but, as it had so obviously resulted from Walker's talk with him, relations between the CO of the 1/6th and his civil colleagues in Ipoh became tense. It was as well that Walker's time in command of the battalion – which earned a Bar to his DSO – was near its end.

In the event he was to leave almost immediately and two months early. An old friend from Burma days, Brigadier Lewis Pugh, had been appointed one of the Deputy Directors of Military Operations at the Ministry of Defence and asked for Walker as one of his senior staff officers.

The job was one of the best Walker could have expected but he did not want to go. It was partly that, as a fighting soldier, he did not want to leave the battlefield and the superb battalion he had created. But he was a realist and recognised that, in order to remain battle-worthy, he himself must now rest. When General Sir John Harding, Colonel of the 6th Gurkhas, who was now the Chief of the Imperial General Staff, had visited the battalion, he had remarked that Walker looked thin and tired and due for a rest. 'After three years on operations,' he said, 'any battalion commander has been under great stress and strain and is due for a rest. You need some leave.' Walker could only agree, remembering he had been out in the jungle for ten days of every month with each company in turn and, as he could not eat Gurkha curry, subsisted largely on Bovril and sweets. He agreed that it was time for a rest.

Soon after, Walker was told that, after his leave, his next job would be as the Senior Staff Officer (Operations and Staff Duties) in Eastern Command, which had its headquarters at Hounslow to the west of London.

It was in November of 1954 that Walker returned to a desk and again began to hanker for active service. Yet his time at Eastern

Command proved to be a busy one. The last major British Army formation in the Middle East – the 3rd Division – was returning and General Sir Francis Festing, the GOC-in-C, Eastern Command, was to be its host. Yet almost as soon as it had arrived, much of the division had to be returned to the Middle East, to Cyprus.

It was not only Cyprus – and, of course, Malaya – that dominated military planning in the Nineteen-Fifties. At the beginning of 1956, immediately after the withdrawal of the last British forces from the Suez Canal Zone, Colonel Nasser, the President of Egypt, nationalised the Canal and one of the most significant watershed years of modern British history had begun.

Soon, orders arrived from the War Office to make contingency plans, then active preparations for the despatch of an expeditionary force to the Middle East – but, clearly, not for Cyprus. To Walker, it seemed a prudent and practical idea to reoccupy the Canal Zone and, as a soldier, he saw no particular obstacle to the operation. But to his amazement and horror he found that, only five years after the end of a major war in Korea and with a major counter-guerrilla campaign in progress in Malaya, the Army could not mount a striking force capable of carrying out a limited operation against Egypt. Not only were there neither enough parachute troops available but no practicable force of transport aircraft or, for the follow-up operations, landing-craft. The assembly of the expeditionary force and the calling-up of reserves was slow and laborious. Not only was this professionally humiliating to a soldier, who prided himself on his readiness for anything, but the delays were obviously giving the Egyptians time to mobilise not only forces but also world opinion.

When the Suez operation finally began in November of that year, only to be abandoned almost immediately under pressure from the United States, Walker was angry. Having taken the crisis that far, it was, in his view, dangerous folly to surrender at this point, for now British threats would never again be taken so seriously. Had he been commanding in Port Said, he told friends, he would have turned Nelson's blind eye to the signals from Whitehall.

Once the expedition had sailed, Eastern Command returned to more routine matters but these were interrupted by an unexpected excitement. Not only had Walker never served in NATO but this was his first experience of soldiering in Europe and the threat of nuclear war which hung over all European defence thinking. Now he was told that he had been chosen as one of about two hundred British and Commonwealth officers who were to watch the British nuclear weapons tests in Australia. Not only was this a fascinating opportunity for any soldier, but it set the seal of approval upon Walker's career, because only those officers earmarked for advancement were to go.

A tight schedule had been prepared so that the spectators would be away from their work for the shortest possible time. But on arrival at the Marralinga Desert Range they were greeted with the news that work on the preparation for the firing had been so slow that their stay might have to be one of many weeks. To hasten the work, the visiting officers were formed into labour gangs and for some forty days set to work preparing the range. The desert below the aiming-point for the airborne nuclear weapons was surrounded by dummy buildings, bridges, military positions, tanks and a variety of equipment, each monitored by strain gauges and cameras. Walker began by digging slit trenches but was soon at work preparing the tanks and fitting gauges and cameras.

When, finally, all was ready on the ground, the wind was blowing from the wrong direction and would have blown the fall-out over farmland. At last the wind changed and the first of the four tests took place. The spectators stood three miles from 'Ground Zero'. They heard the countdown, felt the blast of heat and the scorching wind, then turned to watch the fireball mount into the familiar mushroom cloud.

The spectacle of the explosion was dramatic enough, but even more impressive was the scene on the range when the radio-activity had fallen to a safe level. Solidly-built houses had vanished, bridges were buckled and the tanks Walker had helped prepare were wrecked. But he also noticed that dummy soldiers sheltered in slit-trenches were undamaged and it became at once apparent that nuclear weapons might not be the almighty arbiter of battles that they were supposed to be. Vast explosions, in themselves, had not replaced the brains of military strategists and tacticians.

Walker only saw one of the four tests because Festing's successor, Lieutenant-General Sir Charles Coleman, impatient at his long absence, had ordered him to return. On reaching Hounslow, he gave his opinions on nuclear weaponry and their value and limitations on the battlefield. Coleman was impressed and, for the last six months of his time in Eastern Command, Walker lectured on the effects of nuclear warfare on the battlefield. He did so with added zest because welcome news had reached him. Early in 1957, he was to be promoted to brigadier and return to Malaya to take command of the 99 Gurkha Infantry Brigade Group which he knew was to fight the battle against the most formidable remaining communist terrorists in the southern state of Johore. At that time this command was the most demanding and important of any in the Army. There would, of course, be generals senior to him, but the Battle of Johore was one for a brigade and he would be its brigadier. If it could be won decisively, it should administer the *coup de grâce* to the terrorists in Malaya. He wanted to deliver that blow personally.

South Johore was listed as a priority area towards the end of 1957 and 99 Brigade deployed there as the principal striking force. This consisted of four infantry battalions, with artillery, armoured cars and its own light air support, forming an independent brigade group. For three months this force operated much as it had in other parts of Malaya. Then the new brigadier arrived.

Walker took over command just as the decision to launch a major offensive against the surviving terrorists had been taken. He was asked to choose his own code-name for it and – remembering Churchill's insistence on appropriate, and not flippant, code-names – picked 'Tiger'. The brigade must show 'the cunning, stamina and the offensive spirit of a tiger', he said.

Convinced that such operations could only be successful when the Army and the Police shared responsibility, one of his first moves was to call for a secret briefing from the Special Branch. He wanted to know not only their assessment of the situation but also exactly who their secret agents were and where they were positioned. The marked maps were taken from the safe at police headquarters and he was immediately shocked at the small number of red tabs indicating agents stationed in hostile territory. After hearing the briefing, he was convinced that it would be folly to launch a major offensive based on such flimsy intelligence information.

Walker therefore ordered both his brigade and the police to mount a maximum effort in preparation for the attack. The Special Branch would have to supply the information and, when it was received, he told his battalion commanders, it would be acted upon with such confidence that he must have one hundred per cent success. 'I will not accept failure when you have acted upon Special Branch information,' he told them.

For the battalions, the immediate task was two-fold. Although they were confident in their jungle training, Walker insisted that they make new efforts. Every battalion must have its own ambush range in the jungle and train to a far higher pitch than had hitherto been thought necessary. Commanding officers were each ordered to prepare their own assessments and theories on the coming operation with the object of presenting them at a two-day study period. He wanted each to tell him firstly what they considered their own requirements to be. Did they, for example, hope to deprive the terrorists of their supplies of fresh rice by insisting that villagers receive only cooked rice from controlled cook-houses? Cooked rice would turn sour before it could reach the terrorists and this was therefore an effective sanction. But the provision of these cook-houses and their supervision and guarding was expensive. Did they consider that, in their particular context, the expense was justified?

Then he wanted each to present his own plan for the over-all conduct of the campaign by the brigade. Where did he think the terrorists – probably less than one hundred of them – were in the twelve hundred square miles of jungle in which they hid? How could they best be trapped and killed?

The study period was attended by the executive secretary of the Johore War Executive Committee and Walker was pleased to note that each of his battalion commanders had caught his own enthusiasm and that their assessments were imaginative and vigorous. He was as pleased to note that, in the three months since his arrival, the Special Branch had made successful efforts. One Chinese officer, in particular, was to be as successful as Cyril Keel in 'turning' captured terrorists and setting up a network of informers. Indeed, so many offered their services that, once the operation was under way, he could sometimes attach one informer to each company.

Operation Tiger was not the great set-piece action that was remembered by those who had fought the Japanese. To Walker it was more akin to fighting the Pathans on the North-West Frontier: the enemy were as relatively few and as elusive and the terrain – although densely wooded and humid instead of barren and dry – was as wild and favoured the guerrilla as generously. It was, rather, a series of operations, some forming part of a master plan, others mounted on the spur of the moment – but all flexible, each able to be cancelled or changed instantly at Walker's command.

It was essentially a trap of immense complexity and Walker carried a variety of mobile, interlocking plans in his head. The basis was, of course, his infantry and their supporting arms, notably the artillery and the RAF. Closely tied in with this cutting edge were the Police and their Special Branch and indeed it was they who determined the day-by-day course of operations. Meanwhile the Psychological Warfare Department launched its propaganda campaigns by leaflet, loudhailer, banner and radio across the whole area in the hope of finding a weak link in the resolution of even one of the fugitive guerrillas.

Until a communist terrorist encampment was located and found to be occupied and so require a straightforward infantry assault, the task of the infantry could only be to search the jungle and to lie in ambush. Walker impressed upon them that in order to do this without wasted effort they must not only be the often-quoted combination of poacher and gangster but must live like jungle animals. This was virtually what their quarry had become, he would say. The CTs could smell, say, cigarette smoke or toothpaste for extraordinary distances and their quick eyes never seemed to fail to see the betraying matchstick or sweet-wrapper. The soldiers of 99 Brigade must pass through the jungle, leaving as little trace as phantoms.

Operation Tiger officially began on 1st January, 1958, and Phase One, in which the momentum of the campaign was to build up and pressure on the enemy increase, was to last until the middle of April. The aim was the elimination of ninety-six communist terrorists who were believed to be lurking in one thousand, eight hundred square miles of jungle in South Johore.

When Phase One ended on 14th April, Walker was well pleased with the results. In terms of enemy casualties, these were not spectacular but he had not expected that they would be. Patient patrolling and ambushing and the following up of Special Branch information had resulted in the killing of three CTs, the capture of another and the surrender of three more. One specific military success was that this had resulted in the elimination of one complete terrorist cell. But, more important, 99 Brigade was getting the feel of the jungle and the zest of the hunt and was becoming increasingly sensitive in reacting quickly to Walker's orders.

When Phase Two began on 15th April, 99 Brigade has been built up to four and a half battalions – four Gurkha and half a battalion of the King's Own Scottish Borderers – and very strong Police and Home Guard reinforcements had been made available. The next objective was the destruction of the strongest single terrorist unit in South Johore. This began with the systematic bombing and shelling of the jungle for fifteen days, combined with intensive patrolling, with the aim of driving the enemy from their lairs and into ambushes.

The jungle camps, which were the principal objectives, could be found in two ways: through betrayal by an informer, or by tracking one of the couriers who formed the CTs' means of communication. Sometimes, through an informer, Walker would know where a courier's route would run but he would not know when the courier was due to pass, or know only within a wide margin of time. If the courier could not be followed – and they were expert at concealing their tracks, which might also be obliterated by heavy rain – then he must be killed in the hope that he carried a message of importance, which could be acted upon.

So, throughout the weeks, then months, of Operation Tiger, the ambush became the most demanding task – both physically and mentally – asked of the infantry. They had, of course, rehearsed the varieties of ambush at the Jungle Warfare School and, by this time, most had experienced the reality. But few had had to face the ordeals that now lay ahead.

Now soldiers were told that they must regard an ambush lasting as much as ten days – with two teams alternating in the ambush positions every two hours – as normal. Sometimes these would yield no results and they would be considered successful if they resulted in only a single kill. Walker was insistent that, if he was acting on Special Branch information, the ambush must stay in position until there was no longer the slightest chance of the enemy appearing. This cold resolution brought about what is still regarded as the longest and most demanding ambush in memory.

Information reached Walker that a party of CTs were due to travel a certain jungle trail during the coming month. He therefore ordered the 2/2nd Gurkhas to put an ambush on the trail and accept the fact that the most suitable positions were in a swamp.

The ambush was laid and the Gurkhas melted into the sodden undergrowth to wait. Days passed, then a week, and no furtive figures appeared in their chosen killing ground. After ten days, Walker's staff officers and the commanding officer of the 2nd Gurkhas looked at him quizzically, not yet ready to suggest that he might have been misinformed and that no target would ever present itself. Walker was well aware of the strain he was putting on the ambush party but had not lost confidence in the Special Branch officer's conviction that the enemy would pass this way.

When the ambush had been lying in the swamp for twenty-seven days, the Gurkha colonel finally appealed to Walker. The long wait and the appalling condition of the ground must, he said, have taken the ordeal beyond the power of human endurance. In the interests of his men, he asked Walker to lift the ambush and allow them to rest. Still Walker was adamant. It might already be the longest-lasting ambush ever known but he still believed the enemy would come and, until he himself believed that the enemy would not come, the ambush must stay where it was.

Next day, the Gurkhas, lying exhausted in the swamp, saw a figure move warily along the trail and into the killing ground. Another followed and a third. The Gurkhas fired. Two dropped dead, the third screamed, his rifle shot out of his hand, and turned to run. But, bemused by the shock of the ambush, he surrendered: a prize more valuable to Walker than the two dead terrorists on the trail. Orders went out to lift the ambush.

In other less dramatic ambushes – some of them lasting ten days and more – other terrorists were picked off and, in the most spectacular coup of all, the Special Branch in July persuaded a complete unit of sixteen CTs to surrender. That same month, the 1st Battalion, the Cheshire Regiment, which had relieved the 2/10th Gurkhas, killed the enemy's most important field commander. By the end of August, when Phase Two ended, the terrorists' casualties amounted to twenty-one killed, one captured, five surrendered in action and thirty-one by negotiation with Special Branch. Fifty-eight of the ninety-six had been eliminated.

In scattered but co-ordinated actions such as this, the hundred CTs were hunted and, in ones and twos, killed until, by the middle of August, about seventy had been accounted for. Those killed had mostly been the fighting men, couriers and agents of the communist organisation. But the headquarters and the 'hard core' of the CT organisation still seemed to be intact and, until this could be destroyed, Operation Tiger could not succeed. There was now increased urgency to do this. Events in Johore had immediate impact upon Singapore across the Straits and there elections were due before the end of the year. Upon these elections the future of the Federation and the future possibility of a new Federation of Malaysia – combining Malaya, Singapore and the Borneo territories in a single nation – could depend.

The Communist Regional Headquarters in South Johore was believed to consist of between twenty and thirty terrorists led by its formidable commander, Ah Ann. Cutting the roots and branches of his organisation had been comparatively simple, often a matter of ambushes patiently waiting on trails that the enemy must, sooner or later, use. Now Walker had to track down a small group of highly-skilled guerrillas somewhere within twelve hundred square miles of jungle. To do this he now had only three battalions – the 2nd and 7th Gurkhas and the Cheshire Regiment, with a squadron of the 13/18th Royal Hussars, a mortar troop of the Royal Artillery and a flight of Army Air Corps light aircraft.

For ten years, this regional headquarters had defied all attempts to track it down. But now, with greatly improved information flowing in through the Special Branch, there were some useful clues. Captured documents had given the names, ages and weaponry of twenty-three members of the headquarters. A captured CT courier had told the Special Branch that the headquarters camp was two days march from their position but that he was not sure of the direction or whether it had since moved. An Army Air Corps pilot had sighted smoke rising from the jungle where there was no known habitation. A CT courier was killed by a Gurkha patrol on the bank of a river in spate and it seemed likely that he, too, had come from the headquarters. The northern boundary of the area administered by these headquarters was known. By putting such clues together, an approximate position could be deduced but it was still by jungle standards, a vast area. It had to be searched, trail by trail.

The pressure Walker could now exert was intense. While his battalions pressed through the jungle, the Special Branch was using more subtle means of undermining the terrorists' resolution and surrenders became almost as regular as kills. On one occasion an eight-day chase through the jungle ended with an indecisive fire-fight but, in the confusion, the fleeing terrorists lost their ration packs. The Special Branch advised that the fugitives would have to make for a certain food dump, where emergency rations of elephant meat strips were stored in tins of palm oil. The Cheshires put an ambush on the dump and, sure enough, after a nine-day wait, the enemy appeared as expected, only to escape again when the Bren gun, that should have killed them, jammed. The Cheshires hunted them for some forty miles, then lost the trail. But the effort had not been in vain for, five days later, the starving terrorists emerged from the jungle to surrender.

By the beginning of November, Walker had accounted for ninety of the enemy – his initial aim. The terrorists' organisation had been shattered and South Johore was ready to be declared 'white' – back to peaceful normality. Walker held a Press conference on the 6th and said, 'I will stick my neck out and say that South Johore should be "white" by Christmas.'

The record was impressive. The terrorists' losses had been twenty-two killed in action, one captured, sixteen surrendered, twelve deserted or executed by their own comrades and thirty-nine surrendered by negotiation. But despite this achievement and the congratulations that poured in, Walker was not satisfied.

The most important prize had eluded him. The terrorist leader in South Johore, Ah Ann, his wife and four others had escaped and, it was feared, would be able to vanish across the Straits into Singapore. He was one of the most formidable CTs remaining in Malaya, one of those around whom another insurrection might arise. He was a ruthless man – he had ordered the execution of five of his own men that year – and was known to have large sums of money with him. The chances of his capture seemed remote.

Then Walker's reliance upon the Special Branch bore more fruit. A Chinese officer went to visit Ah Ann's mother on the chance that she might receive word from him, if not a visit. He waited there until a visitor did arrive. It was a messenger from Ah Ann. With the skill that had become routine, the policeman persuaded the messenger that the war was over and that information leading to the arrest of Ah Ann could make him rich for life. The terrorist agreed.

For the mission to his mother, Ah Ann had lent the messenger his own pistol. So it was arranged that the traitor should return it to him, unloaded. This he did and, a moment later, a Police raiding party burst into the camp. Ah Ann raised his empty pistol and squeezed the trigger. Instantly he was shot dead. With him died his wife and another District Committee member. Two others surrendered and the sixth escaped into the jungle – the sole survivor of the terrorists who had dominated South Johore at the beginning of the year. Walker's triumph was complete.

On the last day of 1958, South Johore was declared 'white' and the last major campaign of the Malayan Emergency had ended in the defeat of the insurgents. There was another immediate task for 99 Brigade, the 'mopping up' of other fugitive terrorists in some coastal swamps but, compared with what had gone before, this was to be relatively easy. Praise of Brigadier Walker was the talk of Kuala Lumpur, Singapore and Whitehall.

To Walker himself success had not come as a surprise. Methodically, he had assessed the problem, judged what action had to be taken and then carried this out. He knew what he wanted from his soldiers and he made sure that he got it. His genius for training, inspiring confidence and enthusiasm, which he had applied to the 8th Gurkhas in Burma, had been stretched to embrace a complete brigade. Knowing what he could do with potentially satisfactory material and making it known that those who did not reach his requirements would be sent elsewhere, away from the coming rewards of success, had given him an invincibility.

At the Press conference he had held in November, Walker spoke in a bluff manner, worthy of Montgomery. He reminded his audience that at his last Press conference back in April he had forecast this success. 'Our pursuit of the CTs is going to be ruthless, relentless and remorseless and when we meet them we will hit quick, hit hard and keep on hitting,' he had said. But he did not claim the credit for himself. He praised, in this order, the War Emergency Committee of South Johore, the Special Branch, the information services, the police, the airmen and finally the soldiers themselves, whose tenacity, enthusiasm and skill had, he said, been the greatest factor of all.

In April, Walker's achievement was recognised by his appointment as a Commander of the British Empire. This was followed by a warm congratulatory letter from the Commander-in-Chief, Far East Land Forces, General Sir Richard Hull. Praising Walker's drive and initiative, Hull laid particular stress on the close co-operation he had established with the Police and the civil authorities. Walker seemed to have made a powerful friend here for Hull was clearly about to rise higher in the Army and now he was publicly expressing his satisfaction that the Malayan war had reached its climax with such a brilliant operation and was talking of Walker as 'a very good field commander'.

Now, in one of those sudden and dramatic changes of scene that were to mark his future career, Walker and 99 Brigade were given a task as different from terrorist-hunting as it could possibly be in South-East Asia. Elections were to be held in Singapore and trouble was expected. Although nominally self-governing, the island was still under British rule and, as other old colonies around the world were achieving independence – Cyprus becoming an independent republic within the Commonwealth in this year – its industrious and ambitious Chinese population was becoming restive. Passions were being excited by a young politician named Lee Kwan Yew, whom many regarded as a virtual communist, and there seemed to be a possibility of anti-British rioting, perhaps spilling over into renewed trouble between the Asian racial groups in the city.

Walker was told to assume direct command of internal security throughout Singapore Island, using 99 Brigade as his main force. In all, he would have five battalions at his disposal – three Gurkha and one British from 99 Brigade and the Singapore guard battalion – and he lost no time in assessing their potential by mounting a large-scale exercise. This was a failure, as he had expected it might be.

The task was totally different from that for which the brigade had been forged into so sharp a weapon over the past two years and more. In the jungle his soldiers had been taught to hit quick and hard, to shoot on sight and use maximum force in the shortest possible time. Now, in the streets of Singapore, minimum force would have to be used. His men must avoid the close contact with their adversaries which they had been taught to seek. Before they could open fire, they must warn their opponents by voice, by banner and by bugle call. They must learn to form squares, like those that faced the French cavalry at Waterloo, so as to be ready to hold off rioters attacking from all sides, with a ring of bristling bayonets.

In this first exercise, Walker saw that his men, expert jungle fighters that they were, seemed literally lost in a city. They were slow in throwing a tight cordon round sectors of the city; they blundered into ambushes he had had laid for them; they allowed crowds of soldiers, playing the part of rioters, to stop them by lying down in the road, or to get close enough to try to seize their weapons.

Malaysian occasions. Major-General Walter Walker visits a jungle training area *(above)* before the outbreak of the Brunei revolt. Malaysian gratitude was expressed in June 1965, when Walker *(below)* received the award of Honorary Penglima Mangku Negara—the Malaysian equivalent to knighthood from the Tunku Abdul Rahman in London

Winning hearts and minds. Walker visits Nepalese families with the
2/7th Gurkhas in West Malaysia in 1968

Beryl and Walter Walker meeting a Gurkha soldier's daughter in
school at Sungei Patani, Malaya, in 1961

The five battalion commanders were assembled by Walker. This, he told them, was a poor start, but it would be put right immediately. There was, he had found, no manual of urban internal security operations, nothing beyond inadequate notes and general guidance. So they themselves would write the missing manual. First his officers were told to study Singapore – particularly their own area of the city or island – with intimate care. They must know the road plan and the direction and volume of the traffic flow. They must get to know it so well that it would be unnecessary to glance at a map. Then, when they knew their subject, they were to decide how best it could be controlled.

In a month, 99 Brigade had produced a new military manual, which passed into official use throughout the Army, called *Internal Security in a City*. Here was laid down for the first time the way that soldiers could dominate a city, by occupying key points (usually rooftops), freezing all movement and hovering over it in helicopters. Next, Walker ordered that the theories be put into practice in a series of exercises in which battalions competed with one another in speed and efficiency. At night Singapore was woken by military convoys racing through the empty streets, then spilling out soldiers as they came to a halt. The citizens would look from their windows in the pale dawn light to see Walker's soldiers blocking streets and manning rooftops. He was showing that he could, if need be, dominate Singapore.

Perhaps this show of force acted as a deterrent. Certainly, Lee Kwan Yew was a more responsible politician than many had realised. Anyway, the goal of independence was already in sight. But, whatever the reasons, the elections in Singapore passed off without trouble that could not easily be handled by the police. Lee Kwan Yew was elected Chief Minister and now there was the real possibility that Singapore might unite with Malaya and the former British territories in Borneo to form a Federation of Malaysia.

For Walker himself now came an essential step forward in his profession. He was nominated to attend the Imperial Defence College in London, a training course that would fit him for higher commands and responsibility far beyond that of a defined military operation. He returned home to find himself something of a celebrity in Service circles, if not among the public. The official end of the Malayan Emergency had been declared and the smart, sun-burned figure of Walter Walker was pointed out as the officer who conducted the climactic operation.

In the discussion syndicates at the Imperial Defence College, in Belgravia, it became clear that a new type of senior officer was beginning to emerge. He was a product of the recent troubles in, notably, the Middle East, where British forces had been involved in dangerous military and political crises in Palestine, Egypt, Cyprus and, four years before, the fiasco of the Suez operation. These had produced those known, often admiringly, as 'political soldiers'. It was not that they were prepared to weigh up and act upon political as well as military factors but, in Walker's view, that they were becoming adept at manipulating the political and administrative machinery of Whitehall. This, he thought, was not what soldiers were for. It was for the politicians to make the decisions with the guidance of the civil servants and then it was for them to give a task to their soldiers and trust them to carry it out.

Walker saw some of his fellow-students as becoming more interested in mastering the intrigues of Westminster and Whitehall than in the practical tasks of soldiering. For their part, they sometimes considered Walker too much of a simple soldier. 'When we were given a political and military scenario and told to solve its problems,' one recalled, 'it always seemed that there were none that Walter thought could not be solved with a battalion of Gurkhas.'

But now he was to have more than that at his disposal to solve problems. Three months after he had returned to London, he was told that he was to be promoted to Major-General and to command the 17th Gurkha Infantry Division in Malaya. What pleased him as much, if not more, was that he was also to become the Major-General, the

Brigade of Gurkhas. Now he would not only command all the men of Nepal on active service with the British Army, other than those in Hong Kong and, occasionally, in Britain, but he would also be their father-figure.

For Walter Walker, the Nineteen-Sixties began on a note of high optimism.

5

Whitehall Battlefield

Just as the campaigns fought in the final years of the Roman Empire produced able commanders to lead their legions in rearguard actions and last stands, so the decline of the British Eastern Empire revealed the last of the soldiers who had manned the frontiers, standing like solitary rocks against the flooding tide. Among these few was Walter Walker, but it was not only for military prowess on Asian battlefields that he was of such value.

For almost all of his adult life he had lived among Asians – often, he had nearly died among them – and his knowledge of their attitudes, characteristics and traditions was now to be as valuable as his understanding of their behaviour on the battlefield. Now that the communist challenge in Malaya had been met and defeated, it was hoped that the British withdrawal into Europe and the final transformation of the Old Empire into a Commonwealth, loosely linked by trade and sentiment, could be achieved without further bloodshed. While this task would fall primarily to politicians and diplomatists, it was of the utmost importance that, as far as possible, the final exchanges between the ruled and the rulers should be between friends.

Unconsciously, the Gurkhas had become the keystone of such friendship. They and the British had liked and respected one another for more than a century, taking pride in themselves and in each other. Now the Gurkhas formed the British rearguard in the East and, at this climactic time, Walter Walker was at their head.

Thoughts of war were not dominating most military minds in White-hall or Singapore during the autumn of 1962. In South-East Asia, the dark presence of the Malayan Emergency had rolled back like a mon-soon cloud two years before and now the remnants of the communist guerrillas, who had fought the full weight of Commonwealth military power for twelve years, had fled into the jungle across the frontier in Thailand. There, their presence lent an edge of realism to British and Malayan exercises but otherwise seemed to prove that lasting peace had begun.

Elsewhere, the British found the climate as benign and thoughts of peace had seldom seemed more justified. The Suez disaster, six years before, had finally relieved the British of the self-imposed obli-gation to police the world according to their own rules, although the blandishments of the Conservative Government of Mr Harold Macmil-lan soothed any wounding of pride with the seductive suggestion that the nation had never had it so good.

Other old colonial troubles had been comfortably settled. Two years earlier, the guerrilla wars in Kenya and Cyprus had ended and it hardly seemed to matter that the British had, in effect, won the for-mer and lost the latter because both countries were now indepen-dent beneath the parasol of the Commonwealth. In South Arabia, it was true, the perennial troubles with the hill tribes north of Aden still reminded old India hands of the North-West Frontier but this was of no consequence and was welcomed by some as providing energetic training for young officers. In Aden itself there were, as yet, no signs that the threatened nationalist rebellion need be taken seriously.

None of the three alliances of which Britain was a member – the North Atlantic Treaty Organisation in Europe, the Central Treaty Organisation in the Middle East and the South-East Asia Treaty Organisation – was directly threatened, but the latter did seem on the fringe of troubles. China had been communist for twelve years and North Vietnam for seven and it was assumed that communism was invariably expansionist. It had been contained, then driven back in Malaya but now it was spilling over into Cambodia and Laos and, although neither country was a member of SEATO, Thailand was, and might seem to be threatened. Therefore some limited Commonwealth military intervention here seemed possible to the contingency planners.

Since the French had been expelled from Indo-China the division of North and South Vietnam along the 17th Parallel seemed to be inviting the eventual invasion of the capitalist South by the communist North and President Diem's regime in Saigon was already facing subversion and sabotage by guerrillas. This problem was falling to the Americans, who were supporting the South with military and economic aid and with military advisers. But Britain had been one of the guarantors of the status quo in 1954 and, as such, might be called upon to support political encouragement with military.

Finally there was, to the south, Indonesia. The hundred million people and apparently limitless resources of this immense tropical archipelago had been independent of Dutch colonialism for twelve years and had slipped deeper into that limbo to which military planners consign threatening but vague problems. Had Indonesia been ruled by a recognisable regime of the extreme Left or Right, there would be authoritative guidance from political specialists on means to meet trouble in that quarter. But Indonesia was ruled by an emotional dictator, Sukarno, who seemed to veer wildly between all points of the political compass.

Currently he seemed to be preoccupied in trying to oust the Dutch from their last foothold in Indonesia, West Irian. Since this was the western half of New Guinea and the eastern was an Australian 'trust territory' it seemed that a future threat might emerge here. Since Australia was becoming increasingly sensitive to the real or imagined pressure of expanding Asian populations upon its under-populated spaces – indeed Malaya had been considered the first line of Australian defence since before the Japanese aggression of 1941 – this again raised the spectre of the Yellow Peril. At this time, the British were increasingly confident of their position in South-East Asia and not only because of their undoubted victory over the communists in Malaya, which seemed to have confounded the prophets of doom. Two imaginative innovations – one political, one military – seemed to offer a chance of lasting stability and prosperity.

The first and foremost of these was the formation of Malaysia. The future of Malaya, its Chinese-dominated *entrepôt* Singapore and of the three British territories in Borneo – Sarawak, Brunei and North Borneo – presented so many problems individually that the optimistic hope had grown up that all could be solved at a stroke by amalgamating them. The resulting nation, or federation, would not only allay fears of the increasing power of the Chinese settlers by having a substantial Malay majority but would offer these industrious Chinese a rich and enlarged hinterland to support their trade.

The concept of Malaysia had its critics within and without. Within, the Sultan of Brunei, whose tiny realm was a British protectorate, had no incentive to change his position. With Britain responsible for his external affairs and defence, he could enjoy in security the annual revenue of £12,000,000 produced by his oilfields. As it was in the interest of Britain as much as of himself that the Brunei Shell Petroleum Co. continue to extract 3,500,000 tons of oil each year from the wells of Seria and Miri, there seemed no incentive to share this wealth with his neighbours. Thus, the Sultan was showing little interest in the blandishments of Kuala Lumpur and Whitehall.

Without, President Sukarno saw the formation of Malaysia as the principal obstacle in the way of his dream of Maphilindo. This was the composite name given to the concept of a Malay empire embracing both Indonesia and the Malayan peninsula, all the Borneo territories and stretching east to include the Philippines. This, as Sukarno saw it, would instantly become a giant of some 150,000,000 people, with limitless natural resources, to rival the United States, the Soviet Union or China. To this grand design, the industrial and commercial development of Malaya and Singapore could be seen as the keystone.

But, in 1962, it seemed to British political observers that Sukarno was preoccupied with his ambitions in New Guinea and that the development of Indonesia itself could occupy his attention until such time as Malaysia had evolved into a political and economic entity able to stand on its own, or join with Indonesia on its own terms. Certainly, any political threat from Indonesia faded into insignificance beside the lowering presence of Communism to the north.

A military innovation was designed to provide the Commonwealth with a more efficient and economical defence in South-East Asia. During the Second World War, Lord Mountbatten had fought the Japanese with forces controlled by a single headquarters with himself as Supreme Commander. Since 1945, the three Services had been allowed to command and administer themselves subject only to orders from Whitehall.

They had worked well together under the direction of the High Commissioner and Director of Operations during the Malayan Emergency – particularly when this had been General Templer – but since 1960 they had again gone their own ways. Geography illustrated this separation and the inevitable duplication. The headquarters of the three Services in the Far East were on the island of Singapore. If this is seen as approximately the same size and shape as the Isle of Wight, then Army Headquarters were at Ryde, Naval Headquarters at Ventnor and RAF Headquarters at Freshwater. The two latter Commanders-in-Chief often complained that as joint conferences were held at Phoenix Park, where the Army was established, they always seemed to be on the road, commuting.

Now it was planned that, while the three separate headquarters would remain, there would be a central unified Supreme Headquarters for the control of all land, sea and air operations. This would be manned by all three Services and each would, in turn, provide a Commander-in-Chief. Although the officers in command of each Service would retain his title of Commander-in-Chief he would be subservient to the Commander-in-Chief, Far East, who would, in effect, be

Supreme Commander. Indeed, this innovation was to be made at the behest of the last Supreme Commander, now Admiral of the Fleet Earl Mountbatten, Chief of the Defence Staff in London.

Practical as this plan was, it faced opposition from within all three Services. Not only was there reluctance to accept a reduction in power and status but it was recognised that in the event of war or another emergency, the Commander-in-Chief himself and the Director of Operations he would appoint would be responsible for its conduct, leaving the three Services' Commanders-in-Chief in what could only be a supporting, administrative role.

The new unified command structure was to be established in November, 1962, and the first overall Commander-in-Chief was to be Admiral Sir David Luce, who was currently Commander-in-Chief, Far East Fleet. He, however, was due to become First Sea Lord early in 1963 so, after only a few months, he was to be succeeded by Admiral Sir Varyl Begg, who would come out from the Admiralty, where he had been a successful Vice-Chief of the Naval Staff.

Since the major static force in the theatre was provided by the Army – the strength of the other two Services fluctuated widely with the increased flexibility – the most demanding task in any of the three was undoubtedly that of Commander-in-Chief, Land Forces. For more than two years this had been Lieutenant-General Sir Richard Hull, who was now returning to become Chief of the Imperial General Staff after handing over to Lieutenant-General Sir Nigel Poett.

At this time, the Commander-in-Chief was responsible for an immense variety of military activity. Any troubles between the central Indian Ocean and the China Sea, or China and Australia might involve him. Under his command came several thousand British and Gurkha soldiers and the families that accompanied them. He was responsible for their training and their welfare and, indeed, for the administration of a self-contained community living many thousands of miles from home.

Within the ordered ranks of this martial community was sheathed the principal weapon and instrument of British policy in South-East

Asia. This was the 17th Gurkha Infantry Division, the one major fighting formation outside Europe that was maintained in readiness for war.

This was the force that Major-General Walter Walker had commanded since February, 1961. The pleasure of commander and commanded was mutual on his arrival because one of his three brigades was the 99th which he had trained so thoroughly in the jungle war and when trouble threatened in the streets of Singapore. There were two Gurkha brigades – the other being the 63rd – each composed of two Gurkha battalions and one British. Then based on the newly-built barrack complex at Terendak was the 28 Commonwealth Brigade, the lineal descendant of the Commonwealth Division that had fought in Korea ten years before, composed of a British, an Australian and a New Zealand battalion.

Soon after his arrival, Walker was sent a directive giving him details of policies upon which contingency planning was based. Among these was the directive that he must regard the primary task of 99 Brigade as internal security in Singapore. This seemed reasonable since it was now the most experienced brigade in the arts of riot-control and street-fighting that was available. But he went on to read that there were also political objections to the use of Gurkha soldiers in a SEATO role. For operations in support of the alliance, 28 Commonwealth Brigade was to be used.

Walker was horrified. The limitation on the use of Gurkhas had been imposed by Whitehall, acting through General Hull. Certainly this had no connection with the few restrictions required by the King of Nepal, which were principally concerned with ensuring that Gurkha would not have to fight Gurkha. There had been no objection to Gurkhas fighting alongside British soldiers in Europe during the two world wars and, in recent years, they had fought the Japanese, Indonesians, Chinese and Malays. Why should they now be regarded as unsuitable to fight in Thailand, Laos or Vietnam?

Not only was Walker aghast at this waste of potential, he was unhappy about the use of the Commonwealth Brigade in the 'fire-brigade' role. In emergencies, speed was essential and before this brigade could move, political clearance would have to be received from the Australian and New Zealand governments. The Australians, particularly, did not always agree with British policy in South-East Asia – they were, for example, more ready than the British to appease the Indonesians – and there might well be disastrous delays. The two Gurkha brigades, on the other hand, were a solely British concern and, although Allies would be quickly informed of intentions, they could be moved as fast as transport could carry them.

Walker did not appeal against this directive. Not only was Hull soon leaving to become Chief of the Imperial General Staff but the document had come from his office and it seemed unlikely to Walker that he would have much sympathy with the Gurkha viewpoint. Hull was a cavalryman – he had risen to command his regiment, the 17/21st Lancers in wartime and had later commanded an armoured division – and his post-war record showed that he was unlikely to change his mind at such a moment. Among his appointments since 1945 had been Commandant of the Staff College, Camberley; Director of Staff Duties at the War Office; Chief Army Instructor at the Imperial Defence College and, immediately before becoming Commander-in-Chief, Far East Land Forces, he had been Deputy Chief of the Imperial General Staff. Clearly, he was not an officer who could be persuaded to change policy easily.

There was an alternative. The policy directive was for Walker's eyes; he would write another for his own subordinates. His directive would state the exact opposite.

He wrote: '99 Gurkha Infantry Brigade Group must be trained and prepared to soldier *in any country, in any climate* and *in any sort of war*. To say that the use of British or Gurkha troops in this or that potential theatre of operations may be politically unacceptable is unrealistic. In a *real* emergency, we shall use those formations and units which are nearest, up to strength and acclimatised, both physically and psychologically: those, in fact, which, as General Nathaniel

Forrest said in the War Between the States, can get there "the firstest with the mostest". I want that thought to be behind every activity in which 99 Brigade takes part.'

This directive was written as Hull was handing over to his successor, Lieutenant-General Sir Nigel Poett. He, too, would be unlikely to begin his term as Commander-in-Chief with the changing of a directive that so clearly bore the mark of Whitehall, particularly since he had been Director of Military Operations five years earlier. However, Walker was presented with an opportunity to achieve this with Poett's consent. On taking up his appointment, the new Commander-in-Chief came to stay with Walker at Seremban. It soon became clear that he had not had time to settle down at his headquarters in Singapore to read the files of directives and plans that would soon be occupying his time. So, without reference to the directive he had received from Army Headquarters, Walker handed Poett his own, suggesting that as it was short, he might find time to read and comment upon it during his stay. Poett did so and told Walker that he was in complete agreement with all that he had written.

Before Hull left Singapore, Walker had another cause for alarm. This affected him less as GOC, 17th Gurkha Infantry Division, as in his parallel capacity as Major-General, the Brigade of Gurkhas. This was the senior position within the Brigade and, although it carried no operational responsibilities, it involved every other aspect of their life and work. Walker was responsible for recruiting in Nepal and the negotiations and liaison with the King of Nepal and the Nepalese Army; for the Gurkhas' education and training; for their living conditions and pay; for the selection of their British officers and their own promotion and their morale. In Gurkha affairs, he was now the ultimate authority.

At Seremban, Walker had a separate staff to deal with the Brigade of Gurkhas and it was through this that he received a summons from Hull to visit him at his official residence, Flagstaff House, in Singapore. Personally, they had been on good terms. Hull had written his warm congratulations to Walker on his success when commanding 99 Brigade in 1959 and spoke of him as one of the best field commanders in the British Army. But Hull was five years older than Walker and, within the framework of the Army, their careers could hardly have been more different. So, while feeling no animosity towards one another, the two men were ingrained with different attitudes.

The difference between them was basically that one had been brought up by the British Army, the other by the Indian Army and there was a background of rivalry between the two systems. The British Army was more conscious of social position and might regard Indian Army officers as social climbers who had opted for the rewards which could be earned more quickly in India. Indian Army officers thought that their professional standards were higher than those in the British Army and counted themselves the better soldiers. British officers of the Brigade of Gurkhas found themselves at the extremity of the Indian Army and they had heard themselves and their men mocked as 'Mongolian mercenaries' by boors in both British and Indian regiments. They were therefore, by temperament, inclined to combine the conflicting qualities of supreme self-confidence and wariness.

Hull began by telling Walker that what he had to say was for his ears only. As he would soon be CIGS he thought it right that he should tell Walker what was in his and the Army Council's minds with regard to the Brigade of Gurkhas. He would, of course, welcome Walker's comments as Major-General of the Brigade but the principal object of this meeting was to keep him informed of likely future trends.

The whole British defence effort was under intense political pressure to economise, he said. New weapons systems, the nuclear deterrent and the equipment the Services needed, together with essential improvements in their conditions of service, meant that requirements must be cut to the bone. Lord Mountbatten, as Chief of the Defence Staff, insisted on the most stringent economies and each Service must ruthlessly ask itself what it could sacrifice. The British Army was bound to face cuts and the only outstanding question was where they would fall. Historic infantry regiments might have to amalgamate or disband. What was essential was that the most useful formations – those that, in the long term, would prove most efficient and versatile – must be preserved at the expense of others of slightly more limited value. It was obvious to Walker that a cold eye was now being cast on the future of his beloved Gurkhas.

Then Hull came to the point. It was probable that a straight choice would have to be made between imposing manpower cuts upon the British or the Gurkha infantry. It was therefore the view of the Government, of the Army Council and of himself that the cuts should fall upon the Gurkhas. In the long term it would be the British infantry that would be most important to Britain.

Therefore Whitehall was considering a plan whereby the Brigade of Gurkhas would be reduced from its current official strength of more than ten thousand – the actual figure was nearer sixteen thousand – to four thousand, or one infantry brigade group.

Hull expressed his regret and insisted that he had the highest admiration for Gurkhas. He had not been in favour of the introduction of Gurkha support and specialist units – such as engineers, signals and transport – but he agreed that they were fine infantry soldiers, although, in his view, not necessarily better than the British. Therefore they could not expect to be preserved at the expense of British soldiers.

Walker was appalled. It was not only his personal attachment to the Gurkhas that reacted. He saw the Brigade as the heir to the Indian Army's old role in the precarious balance of power east of Suez. There might no longer be a pressing need to counter a massive Russian threat to the north, or an imminent Chinese invasion from the northeast, but in the sort of campaigns which could be expected – like that so recently won in Malaya or was erupting in South Vietnam – the Gurkhas were in every way the right soldiers to contain and defeat the danger.

Telling Hull that he strongly disagreed with all that he had said, Walker realised that, at the moment, there was little he could do. The briefing had been strictly off-the-record and, even if he could rally support, it was unlikely to have much effect upon the settled views of the future Chief of the Imperial General Staff.

Aghast as he was, the news had by no means taken Walker by surprise. Not only had he watched the development of this policy from several vantage points but a senior officer serving in the Ministry of Defence had, although without direct personal loyalty to the Gurkhas, taken it upon himself to give warnings. Telling Walker of the dangers, he said, 'Hull would never dare treat any British regiment as he is treating the Brigade of Gurkhas. You, their Major-General, will be kept in the dark and never consulted. In the end you will be presented with a *fait accompli*. For God's sake, fight, fight to the bitter end – because if you don't fight, nobody else will.'

However, it did not seem that disaster was imminent. Hull seemed to be thinking aloud, sounding Walker's reactions but none the less determined that this would be the right course to follow when he became CIGS. There would be time to prepare opposition. The key would be the attitude of the King of Nepal and how much he knew of the British Government's intentions. Walker was, of course, in touch with the Military Attaché in Katmandu and a brigadier commanding what were known as Gurkha Lines of Communication – the training depot at Dharan near the Indian border and two recruiting depots – but this was a matter he would have to handle himself. The Major-General, Brigade of Gurkhas, paid an annual visit to Nepal but his next was not due for about a year, towards the end of 1962. Walker decided to restrain his open opposition to the reduction plan until then.

So Poett relieved Hull and Walker set about training his division. He trained it for many different types of war: for internal security operations; for counter-guerrilla fighting; for jungle warfare at several levels of intensity; for a stand-up fight with a well-armed enemy. For jungle warfare, he could use the Jungle Warfare School, which he himself had founded twelve years before, and, in the north, there was a chance of realistic operational training. The Chinese communist guerrilla leader Chin Peng was known to be camped with the remnants of his command just across the Thai border and the Malayan police were glad to have the support of Army patrols.

While in this area, Walker met and made friends with the officer commanding North Malaya Sub-District. This was Brigadier Shelford Bidwell, a Gunner, who was much concerned with the applications of tactical doctrine to training. There was often a natural affinity between officers of the Royal Artillery and the Gurkhas, born of the mutual respect of professionals.

In Bidwell's case this was justified. He was quick-witted, a lucid writer and had a cheerful irreverence in his attitude to the higher strata of command that matched Walker's own. Although an artilleryman, he understood the subtleties of guerrilla warfare and Walker asked him to edit and produce a manual on counter-insurgency operations.

Bidwell saw Walker with a fresh and perceptive eye. He noted the streaks of ruthlessness and puritanism in his character and he admired his qualities as a military teacher. While much Army life had sunk back into a comfortable routine of social and ceremonial activity, Walker would have none of it. He told Bidwell that annual inspections were useless as they encouraged an appearance of efficiency and smartness for the occasion itself and not for long-term military preparedness. 'Instead,' Walker ordered, 'you will ring up a unit at 1800 hours and announce that you will visit them next day.'

His way with exercises kept the division on its toes. He would visit a battalion and, without warning, tell them to mount an immediate assault river-crossing, or set up an ambush system to trap a patrol approaching from a certain direction. Bidwell noticed that Walker not so much imposed his own ideas upon his division as sought and drew out their ideas, combining the best of them with his own, so making each individual aware that he was expected to think as well as to fight.

After Walker had been training his division for a year, Bidwell considered that if it was called upon to fight it would be the best-prepared British field force to go on active service since the British Expeditionary Force of 1914. In 1962, the possibility of involvement in war seemed nearer in South-East Asia. The communist guerrillas in South Vietnam were increasingly active and President Kennedy had heavily strengthened United States commitments there.

Walker had visited the French Army in Indo-China in 1949 and was fascinated by the techniques used by the Viet Minh guerrillas and the increased sophistication their successors, the Vietcong, were introducing. He studied the political and military structure of the Vietcong and their ways of subversion and terrorism. Peasant guerrillas who could operate in the open country of the Mekong Delta as effectively as in the jungle-covered hills of the Central Highlands, who could construct elaborate underground bases and who could conceal their raiding parties for hours at a time under-water, breathing through bamboo tubes, were potential enemies to be reckoned with. That they might have to be reckoned with was a distinct possibility because the Commonwealth Brigade was standing by for service in South Vietnam.

Any such intervention would be under the auspices of SEATO and contingency planning at the new unified headquarters, which had been set up at the end of 1962, was primarily geared to possible external operations within the context of this alliance and to internal security. While the new command structure had met with strong opposition, notably from the Army and the RAF, Walker was pleased with it, for the Malayan Emergency had demonstrated again the importance of flexible inter-Service thinking. In the early days he had seen enough of rivalry between policemen and soldiers to know that bickering over shares of responsibility was playing into the hands of the enemy. Perhaps, as a Gurkha officer commanding 'Mongolian mercenaries' and having few personal friends at the seat of power in Whitehall, Walker was readier than most to judge a man by his abilities rather than the colour of his uniform. In only one belief was he inflexible: that so long as Britain maintained any military presence east of Suez a strong force of Gurkha soldiers was essential.

That this belief was not shared by the British Government became increasingly apparent in private information reaching Walker from London. Friends of the Gurkhas in the House of Commons – such as Sir Jack Smyth and Sir Harry Legge-Bourke – tried to keep their case in the public eye but the Government were not to be drawn. In the House, Mr John Profumo, the War Minister, blandly announced on the same day that the Government was 'not considering tampering with the Gurkhas' and that the question of their future role was 'one of the problems that we are having to look at with regard to the future'.

The signs were sufficiently ominous for Walker to decide to alert the Brigade to the dangers. He therefore drafted an appreciation summing up his convictions and his fears.

This began with a brief history of the Gurkhas' relations with the British Army and of the value of these to Nepal. The 'British Gurkhas' were, he wrote, a stabilising influence in Nepal because they were 'isolated from politics, corruption and intrigue. They are the one stabilising influence in the present delicate situation of friction between India and Nepal and India and China. To remove, or weaken, this prop could only aggravate a potentially explosive situation.'

The Director of Military Operations of the Royal Nepalese Army had, on a visit to Malaya, stressed Nepalese fears over the possible disbandment of the Brigade of Gurkhas, for political and economic reasons.

'All this,' wrote Walker, 'has a direct bearing on the proposal to reduce the Brigade of Gurkhas to half its present strength. Such a decision may well cause serious political repercussions in Nepal; and will inevitably provoke the comment within the Brigade of Gurkhas, and elsewhere, that Britain trades on the loyalty of Gurkhas for as long as it suits her to do so, and then throws them overboard when they become unprofitable or redundant....

'I might well be accused of trespassing outside my province,' he went on. 'But, as Major-General, Brigade of Gurkhas, I owe a separate and immutable allegiance to Gurkha soldiers and Gurkha pensioners, and to their country, and I am not prepared to betray the trust....'

Walker then examined various measures which could be taken within the Brigade to minimise the effect of the proposed cuts. The language he used was tough and uncompromising, at one point writing of 'a disastrous effect on the morale of officers and men.... I could not hold myself responsible for what might come to pass.'

And he returned again to the theme of betrayal: 'It has taken time to build up the confidence and trust of the Gurkha soldier in the British Army and it has now reached its zenith. The news that the British Government intends to ditch them ... will come as a great shock and will be an extreme test of their loyalty. To argue that the Brigade of Gurkhas must make its contribution to any further run-down of the British Army is unlikely to have much impression on officers and men, because the previous Gurkha brigade was cut from ten Regiments to four in 1948 (on Indian independence) and, in their eyes, they should not be expected to suffer yet another cut....

'It is going to be difficult to give them convincing reasons why the British Government is now throwing away an Ace card at a time when British manpower is short and some more potent argument than the overseas balance of payments, redundancy or the political difficulties to using them in cold or limited war, will have to be devised.'

He concluded with a final plea for the retention of the Gurkhas as the most effective troops for use in South-East Asia.

'Only Asian troops will be acceptable in an Asian country,' he argued, pointing out that Singapore would no longer accept white troops for internal security duties and that, while Gurkhas had fought or were fighting other Asian races, they themselves were essentially 'neutral, acceptable to all shades of political opinion'.

'No political objections are raised when it suits us to use Gurkhas,' he summed up. 'Surely, therefore, it is sensible and in our interests to retain a Brigade Group of Gurkhas in Malaya. Malaya is the last bulwark against Communism in this part of the world. If Malaya falls, the situation in South-East Asia becomes irretrievable.'

Walker was preparing for his second official visit to Nepal with two objectives. One, of course, was to talk to his British officers and inspect their administrative and recruiting organisations, and to talk with the Nepalese military commanders and make his report on British Army Gurkhas to the King. The other was to discover what the Nepalese knew about the British Government's intentions and what their reactions were. An important part of the tour was a series of visits to Gurkha pensioners in the hill villages from which most of the recruits came. These could often only be reached on foot and the Major-General, accompanied by a small British and Gurkha staff and porters, would for its duration, which might be as much as ten days, be out of touch with the outside world because wireless equipment could not then be part of the baggage they carried over the hills.

Before making final arrangements, Walker checked with Headquarters, Far East Land Forces, to make certain that there were no political or military storm-clouds on the horizon that would make it prudent to postpone the tour. He was not only assured that there were none but, he was told, General Poett together with the Commander-in-Chief, Far East Fleet, Admiral Sir Desmond Dreyer, and the Commander-in-Chief, Far East Air Force, Air Marshal Sir Hector McGregor, would also be away. They were flying to the Philippines to watch a demonstration of American naval and air fire-power in the China Sea. Admiral Luce was remaining in Singapore with his Chief of Staff, Major-General Brian Wyldbore-Smith, but there was no hint of trouble that might involve the Far East Command.

So, with great satisfaction, Walker settled down to draft his report for the King of Nepal. This was not to include any hint of his anxieties over the future of the Gurkhas with the British Army; that was a question to be raised personally in Katmandu.

Accompanied by the brigade major and his ADC, Walker flew to Katmandu via Calcutta early in December. These had become pleasant but routine occasions, for British relations with Nepal were excellent. The Nepalese Government laid down that Gurkha soldiers must not be used against other Nepalese, against Hindus or against civilians but their attitude was flexible and they raised no objection to their involvement in internal security operations in Singapore. They were equally obliging towards India, which, had inherited more than half the old Indian Army's Gurkha battalions, which had since been strengthened and expanded to nearly fifty battalions. When China complained to Nepal at the use of Gurkhas by the Indian Army to defend the northern borders, they were rebuffed.

These good relations were based on mutual trust established by the old Indian Army and continued by the new and by the contracted British Army. While day-to-day liaison was provided by the British military attaché in Katmandu and other officers in Nepal, the main point of contact was the Major-General's official report, personally delivered, and the subsequent informal audience with the King. This was once a familiar scene in the vanished British Empire: a British general paying his respects to an Asian potentate, each aware of the manifold motives behind the polite formalities.

King Mahendra received Walker soon after his arrival, together with the British Ambassador. After the greetings, Walker read out his report, which was short and optimistic. Couched in courtly language, this began with an account of the activities planned for a Gurkha battalion group training in England – an idea which had, in fact, been actively supported by the new CIGS.

He went on to describe in outline the various Gurkha activities, from counter-terrorist patrolling in the jungle of North Malaya to the training of Gurkha girls as nurses with a view to establishing a Gurkha Nursing Service. He mentioned the fall in the incidence of tuberculosis among Gurkhas who were, after coming down from their native mountains, particularly vulnerable to the disease. And he wound up, saying, 'I can assure Your Majesty that the British Army continues to look after the interests and welfare of Your subjects....'

There was nothing here to displease Whitehall, when a copy of the report was read there, nothing to ruffle diplomatic or military protocol and no hint of what was now to come.

The King indicated that the informal audience could now begin and invited Walker to raise any question he wished. He came straight to the point. Had the British Government consulted the Government of Nepal about its proposals to drastically reduce the number of Gurkhas serving with the British Army? The King replied that they had not been consulted and urged Walker to continue. The British ambassador showed signs of embarrassment.

In that case, said Walker, Nepal might expect to be presented with a *fait accompli*. The position, as he understood it, was as follows. The Prime Minister of Great Britain might have asked the Foreign Secretary whether there was any reason why the Brigade of Gurkhas should not accept a share of the coming cuts in defence expenditure.

The Foreign Secretary might have reminded him of Nepal's economic dependence upon the British Army which gave employment to many thousands of its young men, who remitted some of their pay to their families and finally retired to their villages on pension. Severe cuts in Gurkha manpower would have a serious effect upon the stability of Nepal, an old and valued ally whose friendship was now as important as ever. In Walker's own view, the Gurkhas had proved this very recently for it was they who had saved Malaya from communist domination.

Then the Prime Minister might have consulted his Secretary of State for Defence and the Chiefs of Staff. They, as military men, might not have understood the wider implications and have insisted not only that the Gurkhas accept cuts, as any other arm of the Services would have to, but that the strength of the Gurkhas should be reduced before that of the British Army.

If the King wished to avoid such a situation he would be well advised to make his voice heard in London.

Clearly disturbed, the King thanked Walker and said that he would consider what action to take. Then Walker and an apparently discomforted ambassador took their leave.

The Borneo battlefield. A rifleman of the 1/7th Gurkhas on patrol in Sarawak in 1964

A jungle fort. The strongly defended position at Bukit Knuckle in Sarawak
was one of many patrol bases along the border of Kalimantan

Complaints are heard. A sergeant of the Argyll and Sutherland Highlanders shows Walker and Sir Solly Zuckerman, the British Government's Chief
Scientific Adviser, the field rations that had been found unsuitable for jungle
operations

But it was at the British Embassy that Walker now met a powerful ally. Mr Henry E. Stebbins, the United States Ambassador, had come to regard the employment of Gurkhas by the British Army as one of the major stabilising influences – economic, political and military – in the life of Nepal. He was appalled when Walker told him of the British Government's intentions and invited him to his house for further private talks.

Stebbins listened to Walker's case for the preservation of the Brigade of Gurkhas and found himself in total agreement. Then he put forward his own arguments which were those he would express to Washington.

Nepal was badly in need of foreign exchange and the Gurkhas' remittances and pensions formed by far the largest source of this. Thus to reduce, or possibly disband, the Brigade would have a serious effect on the country's economy. As the United States was giving Nepal aid amounting to between three and four million dollars each year, this was obviously a matter of concern to him.

Politically, the psychological barriers to the progress of Nepal were its dependence upon India and its fear of China. The recruitment of Gurkhas by the British gave the country a self-reliance, as Britain was considered a powerful friend and ally. If recruiting by Britain was stopped then it would certainly be continued by India, so increasing the dominance of New Delhi over Katmandu. Another important political factor was the number of retired Gurkha soldiers who returned to their own villages and settled down, becoming a source of stability and potential leaders of their communities.

Militarily, the existence of a large reserve of well-trained ex-soldiers was a great asset to Nepal. Their skills were not all strictly military and their training in technology, transport and administration made them a national asset in a wide field.

Walker, of course, agreed with these views. 'You and I are on the same side,' he told Stebbins and they agreed that it was extraordinary that the British Government had consulted neither Nepal nor the United States while planning to take a step with such far-reaching consequences. But now Stebbins promised Walker he would do everything he could to support him and that pressure might be brought to bear on the British Government.

This he did. After an audience with the King to discuss Walker's warning and talks with the Commander-in-Chief of the Royal Nepalese Army, Stebbins reported to the State Department in Washington, giving details of his talks with Walker and setting out his own strong views. Another fuse had been lit.

Before leaving Katmandu, Walker wrote a guarded report on his meeting with the King to Admiral Luce. He received a letter from the King's Military Secretary, beginning, 'By command of His Gracious Majesty I have the honour and pleasure of conveying His Majesty's blessed THANKS to you for submitting before him the Annual Report of the Major-General, Brigade of Gurkhas, for the year 1962 personally.' Satisfied that he had done all he could, Walker left for his tour up-country.

Such treks on foot through the hills of Nepal were to most Gurkha officers idyllic. There had always been a strong tie of affection between the British and the Gurkhas and to the former it was heartwarming and moving to be welcomed into simple Gurkha villages and greeted with smart salutes by grizzled pensioners wearing their campaign medals, agog to exchange memories of Burma, the Western Desert or Monte Cassino.

Although out of touch with Katmandu, Seremban and Singapore, Walker felt relaxed. At his audience with the King he had done what he saw as his duty. To him, it was more important to make every possible effort to save the Brigade of Gurkhas from crippling reductions and probably eventual disbandment than to abide by protocol.

The British Government's plans seemed not only a betrayal of the Gurkha soldiers themselves and their loyalty and trust but a betrayal of the whole British Army which might be called upon to fight in the East without the essential experience and support that only the Brigade of Gurkhas could provide. But, in speaking his mind, Walker could not fail to be aware that Whitehall would consider him not only guilty of a breach of protocol but of trust because it would never be known that he had had other sources of information than Hull. The CGS would probably find him guilty of a breach of confidence, disobeying orders and even insubordination. On the other hand he was doing what he considered right – indeed, he was taking the only possible and honourable course – at the risk of his own career. In Walker's view, it was the British Government and the Army Council that was guilty.

Meanwhile, accompanied by his small staff and the brigadier commanding the depot at Dharan, he could enjoy the company of Gurkhas and the keen mountain air, refreshing after the humidity of Malaya.

The dawn of 8th December was particularly beautiful. As the sun rose over Mount Everest, the five British officers assembled for breakfast at their hillside camp and to listen to the news on the ADC's transistor radio. The bulletin began with an announcement so unexpected and so laden with possibilities of disaster that for a moment they were speechless.

Revolt had broken out in Brunei.

Walker broke the silence. 'I'm in the wrong place,' he said.

No hint of trouble in Brunei had reached Walker. Nor could it have reached the three Services' Commanders-in-Chief because they had gone to the Philippines. If rebellion could suddenly erupt without warning in the heart of what was regarded as a British sphere of influence, what else might happen? Was South-East Asia about to be swept by insurrection?

For such a moment, Walker had spent his life preparing himself and his soldiers. Now that the moment had come, he was more than two thousand miles away from the division which he should at this moment be leading into action. Moreover, he was in the desperate position of being able to receive news from the outside world but totally unable to communicate himself. Not only had he no wireless transmitter but he was three days march from the nearest airstrip. Even had he been closer it would have been of no avail because the Dakota aircraft that was to collect them, to fly them back to Katmandu, was not due until then: it was not even certain that, weather and the serviceability of the aircraft would enable it to arrive on schedule. So, heavy with frustration and foreboding, the party struck camp and began the first step of the journey to Brunei on foot.

The rebellion that took the Far East Command by surprise was not one of the communist-inspired insurrections which they had expected. It was primarily a domestic revolution aimed at political and social reform in the Sultanate of Brunei but, far from wishing to replace the Sultan with a left-wing junta, it aimed to re-create the empire of Brunei which ruled much of Borneo and adjacent islands in the sixteenth century. The rebels did enjoy some external help but it was not from communists nor from the Chinese. Indeed, fear of Chinese domination made the rebels as hostile to the idea of Malaysia as the Indonesians.

Brunei had been a British protectorate since 1888 and was the smallest of the four states that made up the island of Borneo. More than two-thirds of the island's 287,000 square miles had belonged to the Dutch and had been inherited by Indonesia and renamed Kalimantan. The remainder, stretching about one thousand miles east to west and running from the crest of a watershed mountain chain to the northern coast, was dominated by the British. The largest state, Sarawak, with a population of nine hundred thousand, had been granted by the Sultan of Brunei to Sir James Brooke, a British merchant-adventurer in 1841, and he had sired the remarkable dynasty of 'white rajahs'. After the Japanese occupation, the Brookes ceded Sarawak to Britain and it became a Crown colony, which it was to remain until the formation of Malaysia.

Sarawak occupied the western half of the northern seaboard and North Borneo, with six hundred thousand inhabitants, lay to the east. This, also ruled by Brunei, had later become notorious for Dyak piracy until it became a British colony in 1891. This, too, was to join Malaysia, under the name of Sabah.

Between them lay what was left of the Sultanate of Brunei. This ran inland on either side of major rivers in two deep salients into

Sarawak and North Borneo. It covered only 2,226 square miles and of its eighty-five thousand people, just over half were Malays, a quarter Chinese and the rest the native Dyaks of Borneo.

Originally, Brunei, like the other states, depended upon agriculture and this was still its people's principal occupation. Pepper plantations lined river banks and up-country, on the slopes of the mountains, virgin or primary, jungle, offered an almost inexhaustible source of fine timber. In 1929, oil was discovered on the coast and the concession to the Shell Petroleum Company at Seria provided the Sultan with his immense income.

Signs of this wealth were apparent in Brunei Town, beside the Brunei river ten miles from the sea. A huge concrete mosque, its dome sheathed with gold leaf, dominated the capital and, just outside the city, the Sultan's sumptuous Istana stood on a bluff overlooking the river. Otherwise, Brunei was a commonplace market town and minor seaport, less well-ordered and clean than towns in Sarawak and North Borneo where the Chinese were predominant. For visitors, the sights of the city were the mosque and, beside it, the sprawl of rickety bamboo and attap dwellings built on stilts driven into the muddy bed of the river. Unequal division of wealth was as startlingly illustrated here as anywhere in South-East Asia. That this was a likely setting for revolution had long been a vague thought at the back of colonial administrators' minds – but there the thought had remained.

A revolt had broken out in 1953 and it was easily put down, but its lessons had not been learned. Reforms did not follow and there was still justifiable resentment at inadequate housing, education, health services and water supply for the people on the one hand, and the inefficiency, corruption and riches of the small ruling class on the other. The leaders of the revolt – many of them intelligent idealists – had returned to their homes and, since the Special Branch of the Brunei Police had been allowed to dwindle and degenerate, were able to begin planning for another attempt.

The Sultan, Sir Omar Ali Saifuddin, was not unenlightened yet he could see no alternative to a despotic regime. In a half-hearted attempt to present at least a show of democracy – Sir Omar was a passionate admirer of Sir Winston Churchill and perhaps hoped to emulate the near-dictatorship that he had exercised in Britain during the Second World War – the Sultan agreed to the first-ever elections to a Legislative Council for which he himself would nominate more than half the members. The election was held in September, 1962, and all the seats contested were won by the reformist People's Party – the *Partai Ra'ayat* – which was in favour of joining Malaysia provided that this was preceded by the unification of Brunei, Sarawak and North Borneo with their own Sultan as the constitutional ruler. The new state, with a population of more than one and a half million, would then be in a position to resist domination by Malaya or Singapore, Malay administrators and Chinese merchants.

Like other frustrated political movements, the *Partai Ra'ayat* bred a militant – indeed, military – wing. This was the North Kalimantan National Army, otherwise the *Tentera Nasional Kalimantan Utara* or the TNKU. This saw itself as an 'anti-colonialist' liberation movement, another EOKA, Mau Mau or IRA. Its sympathies lay more with the Indonesians, who had violently cut their ties with the Netherlands, rather than with Malaya and Singapore which seemed still to be dominated by the British imperialists. This tendency was encouraged by its leaders. One, A. M. Azahari, a thirty-four-year-old political agitator of Malay and Arab descent, who had lived in Indonesia and was in touch with Indonesian intelligence agents, was the leader of the TNKU. As commander of the armed force he was recruiting he appointed Yassin Affendi and among his officers were several who had been trained in clandestine warfare in Indonesian Kalimantan and in Jakarta. Towards the end of 1962 the TNKU could muster some four thousand men, a few modern weapons and about a thousand shot-guns. It was not much of a military force but enough, they thought, to stage a *coup d'état*.

In November, no hint of trouble in Brunei had reached Headquarters, Far East, in Singapore. As a result of the 1953 rising, there was a contingency plan for the reinforcement of the Brunei Police by one infantry company, but that was all.

The first hint of trouble came in early November. The new British Resident in the Fifth Division of Sarawak, which adjoined Brunei, was Mr Richard Morris, an Australian. He arrived at the Residency in Limbang, a small town on the Brunei river about ten miles upstream from the capital. Soon after he arrived on 8th November, he was visited by an old friend from Brunei who told him that 'something is blowing up'. This led Morris to call in two of his District Officers and, after sifting scraps of rumour, decide to warn the authorities in Kuching, the capital of Sarawak, that 'some political intrigue seemed to be going on'. As a result, Special Branch officers visited Limbang but their enquiries and searches only discovered some illegal uniforms with TNKU badges.

On the 23rd, Morris received a report that an insurrection was planned in Brunei but that it would not take place before 19th December. This was passed to Kuching and from there to Singapore and Kuala Lumpur. As a result Mr Claude Fenner, the Inspector-General of the Malayan Police, flew to Sarawak to investigate, found no more firm evidence but, on his way back through Singapore, gave Admiral Luce what was described as 'a vague warning'. The Commander-in-Chief asked his Chief of Staff, General Wyldbore-Smith, to look out the contingency plans for Brunei and bring them up to date in the light of Fenner's report.

The contingency plan which was returned, amended, to the file, was known as Plan Ale. This involved the flying to Brunei of a small headquarters, two rifle companies with light-scale equipment, small detachments of the Royal Engineers and Royal Signals and one Military Intelligence officer. There was no naval involvement and no requirement for the RAF after the initial fly-in and the only subsequent move mentioned was possible reinforcement by a third company. Certainly, it was thought that there was no need to declare any alert in Brunei, put any forces in the first state of readiness, or inform the three absent Commanders-in-Chief, or attempt to get in touch with General Walker in Nepal.

On 6th December, Morris heard that the rebellion was now due to start on the 8th. Next day, more information reached the Resident of the Fourth Division at Miri, a small town and an oilfield on the coast of Sarawak, some twenty miles west of the Brunei border. This was Mr John Fisher, who had joined Rajah Brooke's administrative service as a cadet of nineteen and now, in his late fifties, was one of the most experienced observers of the moods and political undercurrents in Borneo. He was also particularly aware of the possibilities of trouble because his brother-in-law was commanding the main British strategic force in the Far East and he and his wife Ruth often stayed with the Walkers when they were visiting Malaya. Early on the morning of 7th December, Fisher was warned by one of his most reliable contacts in the district that rebellion was imminent. He at once informed the Sarawak Government and the warning was passed to the Commissioners of Police in Brunei and North Borneo. Police were put on full alert throughout the country and, that same day, a platoon of the Sarawak Field Force police was flown to Miri. Yet, even then, no action was taken by the Commander-in-Chief, Far East.

The rebellion broke out at two o'clock on the morning of the following day. The first signals arriving from Brunei at Headquarters, Far East, reported rebel attacks on police stations, the Sultan's Istana, the Prime Minister's house and the power station and that another rebel force was approaching the capital by water. Only then did Admiral Luce take action and the order 'Ale Yellow' was given, putting the small force of two companies at forty-eight hours notice to move. As Walker had forecast, these were not from the Commonwealth Brigade but from the 99 Gurkha Infantry Brigade – and Gurkhas at that – which was near the airfields at Singapore.

More news came through from Brunei. Thanks to the advance warning, most of the rebel attacks in the city had been repulsed but they had captured the power station and cut off the electricity supply. Far more serious was the situation at Seria, where the rebels had captured the police station and were dominating the oilfields. Little more was known at the time but the rebels had, in fact, attacked many police stations throughout Brunei, the Fifth Division of Sarawak and in the western edge of North Borneo. Miri was still in Government hands but Limbang had fallen to the rebels and Morris and his wife were among those taken prisoner.

Nine hours after the 'Ale Yellow' alert, Admiral Luce ordered 'Ale Red' for immediate action and two companies of the 1/2nd Gurkhas took the road to the RAF airfields at Changi and Seletar.

The RAF was in an even lower state of readiness than the Army – indeed neither had been ordered to make any preparations until the early hours of that morning – but there were usually some transport aircraft available. This morning there were by chance, in addition to old Hastings medium-range transports, four large aircraft: one Britannia and three Beverleys. It was planned that they should fly to the island of Labuan in Brunei Bay, from which they could be flown in light aircraft to Brunei airfield if this had not been lost to the rebels.

It was vital that the troops should reach Brunei before dark. Not only would it be far more dangerous to enter an unfamiliar, enemy-held town at night but a show of force would only be effective if it could be seen. The main preoccupation of the commander of 99 Brigade, Brigadier A. G. ('Pat') Patterson, was to see Ale Force – as the expedition was named – airborne. To his disgust he was ordered to remain in Singapore with his brigade headquarters for, so sanguine was the Commander-in-Chief's view of the crisis, that the commanding officer of the 1/2nd Gurkhas was not ordered to Brunei. Ale Force was to be led by his second-in-command, a major.

But Patterson and Brigadier Jack Glennie, the Brigadier General Staff from Far East Headquarters, hurried to Seletar in the hope of seeing the troops before they left. He need not have hurried. To his horror the scene there was one of tranquillity. The RAF staff was going through all the laborious procedures of passenger embarkation. Not only was the name of each Gurkha being entered on the manifests – Gurkha names are notoriously difficult to spell to those not speaking Gurkhali – but each man was being weighed with his weapon and equipment.

The sight enraged Patterson. Vital time was being lost not only through unnecessary formalities but through stupidity. He asked an RAF officer why it was necessary to weigh the Gurkhas when it was obvious that their average weight was less than that of the British and the number of seats in the aircraft was known. He was told that these were the regulations. Patterson seized the passenger manifest, flung it to the ground and ordered the RAF to take off without delay.

When the Beverleys lumbered into the sky from Seletar and the Britannia from Changi that afternoon it was too late. Inevitably the troops would now be unable to reach Brunei Town before dark. In one report on the operation, it was recorded: 'It was vital to get troops into Brunei Town before darkness on 8th December. They would then have been able to dominate the rooftops of the main square and initial casualties would have been saved. This could have been done if there had been some streamlined process of getting at least one Beverley, loaded with troops, off the ground. However, both Army and RAF staffs at the airfields in Singapore were unaware of the tactical urgency of the problem and no attempt was made to "cut corners" until the joint intervention of the BGS, GHQ and Cmnd. 99 Bde.'

Once Ale Force was airborne, sloth gave way to a brave display of initiative. While airborne, it was learned that the rebels had not yet reached Brunei airfield so the three Beverleys were diverted to land there, while the Britannia headed for Labuan as planned. The flight of seven hundred and fifty miles ended some twenty hours after the rebellion had broken out when the Beverleys touched down at Brunei airfield and the Gurkhas began their advance on the town.

General Poett had arrived back in Singapore seventeen hours after the rebellion had broken out but, even then, Brigadier Patterson was not allowed to take charge of his brigade's operations in Brunei. Instead, the Commanding officer of the 1/2nd Gurkhas flew out to take over from his second-in-command and, on the evening of 9th December, Brigadier Glennie followed with another small headquarters to relieve him. It was not until Patterson had demanded that Poett allow him to command his own troops in action that he flew out on the 10th.

In Brunei, the action was swift and decisive. The Gurkhas had moved into the capital and fought a series of sharp actions with rebels, in which they suffered six casualties, two of them fatal. The Sultan was found unharmed and brought from the Istana to police headquarters for greater safety. An attempt to reach Seria by road ran into strong opposition and, after some initial success, was ordered to return to Brunei Town to meet a threat of rebel counter-attack on its centre and on the airfield.

Meanwhile, Fisher, realising how widespread was the rebellion and how thinly-spread the available soldiers and police, took a dramatic step. He had many friends among the tribes up-country and, on the 9th, he called upon them for help. The longhouses where they lived on the banks of the Baram river were out of touch by modern communications, so he resorted to the traditional and sent a boat carrying the Red Feather of War – the ancient summons to tribal warriors – up the Baram.

To his influence was added that of Mr Tom Harrisson, the Curator of the Sarawak Museum at Kuching, who had been flown to Brunei as a Government observer. This was a remarkable man. Thirty years before, when an undergraduate, he had come to Borneo with a university expedition. Back in England he had founded the first opinion poll organisation but had returned to Borneo in 1944 to raise the tribes against the Japanese. Since then he had become a leading authority on the many aspects – including the ecology and the archaeology – of Borneo. What was now particularly useful was that his wartime exploits had been among the hardy Kelabit people who lived in the mountains inland from Brunei. Now he, too, summoned his old friends to his aid.

Rallying to both calls, the tribesmen came down-river in their hundreds. Those without their own weapons were armed with shot-guns and they were formed into company-strong groups led by British civilians, the whole being commanded by Harrisson. What came to be known as Harrisson's Force grew to a strength of nearly two thousand and played a major part in containing the rebellion and, later, in cutting the escape routes of fugitive rebels.

As important was the detailed intelligence both Harrisson and Fisher were able to provide about jungle trails for, apart from the roads linking Brunei Town with the oilfields, these and rivers were the only surface communications.

Reinforcements were now flowing in from Singapore, using Labuan as their advance base. The 1/2nd Gurkhas were brought to battalion strength and, early on 10th December, the next 'spearhead battalion', the Queen's Own Highlanders, began arriving at Brunei airfield. The immediate threat to the capital had passed and when Brigadier Patterson arrived to take command he was faced with the task of recapturing towns and villages which had been in rebel hands for nearly three days. The most important of these were Seria and Limbang and in both the rebels were believed to be holding British hostages.

The shortcomings of British defence planning in Singapore and the ossification that had overcome its organisation were now to be redeemed by two gallant and imaginative operations.

On 10th December, a second attempt was made to relieve Seria. The eighty-mile journey by road, through patches of dense jungle, had been shown by the first attempt to be fraught with the danger of ambush which must by now have increased. Lacking naval support, the only alternative was assault from the air. First, a reconnaissance flight was made over the oilfields. Rebel flags were flying over the Shell complex – the office blocks, the club buildings and the sumptuous houses of the executives set among lawns and flowering trees – and the whole six-mile stretch of coast seemed firmly in enemy hands. However, it could be seen from the air that there might be a clear landing zone for light aircraft to the west of Seria and east of the town; the runway of the airfield at Anduki, although rebel-held, had not been obstructed.

A small air armada was assembled at Brunei airfield – one Beverley, five small Twin Pioneers and a Beaver reconnaissance aircraft of the Army Air Corps – to embark one hundred men of the Queen's Own Highlanders. Near Seria, the force divided to land ten miles apart; the Twin Pioneers on rough grass to the west of the town, the Beverley on Anduki airfield. Both assaults achieved complete surprise.

The Twin Pioneers survived rough landings and their soldiers were able to disembark and seize a police station two miles away without loss. The only fight was for the Telecommunications Centre in Seria, where ten rebels were captured.

The Beverley landed with its soldiers standing by the doors ready to leap out when it touched down. As they did so, the pilot opened the throttle and the great aircraft soared away under fire from the airfield buildings. The Highlanders fought a sharp action for the control tower and the airfield was quickly captured.

Reinforcements, arriving from Singapore via Brunei, were flown in but it was not until the 12th that the Seria police station was finally stormed and forty-eight European hostages rescued.

Meanwhile, the recapture of Limbang presented an even more difficult problem for there was no possible air landing-zone near the town and no naval craft available for an assault from the river.

Among the reinforcements that had flown into Brunei were Royal Marines of 3 Commando Brigade, which was based on Singapore island and on one of the two commando ships, *Albion* or *Bulwark*. By nightfall on 11th December, eighty-nine Marines of 42 Commando had arrived and were embarked in two unarmoured cargo, lighters fitted with ramps like landing-craft. Limbang was known to be held in strength by the enemy, so speed and boldness being the essential ingredients of a surprise attack, it was decided to launch the attack up-river during the night.

Ashore in Limbang, Morris, the Resident, his wife and six other hostages had been expecting an attempt at rescue. But knowing how strongly the town was held – by about one hundred and fifty rebels it was thought – they feared not only that the attempt might be ambushed but that they themselves would be killed by their captors. In fact, the rebels were making plans to hang Morris that same afternoon, whether or not a rescue attempt was made.

At first light on 13th December, Morris heard the throb of marine engines out on the river and through a window could see what appeared to be two landing-craft approaching the town. At that moment, machine-gun fire opened and the assault began. It was quickly over but, in routing the rebels, killing fifteen and capturing eight, the British lost five killed and eight wounded.

By the following day, a series of similar, but smaller-scale operations, had recaptured the lost towns and villages and the back of the rebellion had been broken.

On the 17th, Walker finally reached Singapore via Katmandu and Calcutta. He hurried to Far East Headquarters and was given an immediate briefing by Poett. The rebellion had been held and broken, he was told, and fears of any spread across Sarawak and North Borneo had proved unfounded. Of the original four thousand rebels, forty had been killed and some three thousand four hundred captured. The remainder had fled into the jungle and might be trying to reach the Indonesian frontier. Of the two principal leaders, Azahari had been away in the Philippines throughout the revolt, and Yassin Affendi was among the fugitives.

For the mopping-up operations a considerable force was now available. The original two battalions had been reinforced by the whole of 42 Commando and the 1st Green Jackets had landed at Miri from the cruiser *Tiger*. The commando ship *Albion* had been making for Miri, where she was to disembark 40 Commando, but she had been diverted to Kuching, the capital of Sarawak, as a move to pre-empt trouble among the Chinese. In all the towns of Sarawak there were known to be cells of the Clandestine Communist Organisation – the CCO – and although there had been no disturbances, these were openly sympathetic towards the Brunei rebels. Kuching itself lay only thirty miles from the Indonesian border and was obviously a potential flashpoint. As a precaution, the Sarawak Police had arrested about fifty known Communist leaders and had closed the communist Press. On the 12th, Admiral Luce had flown a battery of the Royal Artillery, without its guns, to Kuching to help with internal security. On the 14th, they had been reinforced by most of 40 Commando.

The *Albion* had then steamed eastward to land the remaining Commando company near Seria and to disembark one of her two helicopter squadrons at Brunei.

The arrival of the Whirlwind helicopters of 846 Squadron, Fleet Air Arm, and the presence off-shore of the Wessex helicopters of 845 Squadron added a new dimension to operations. Now it was no longer essential to find makeshift airstrips for transport aircraft and attempt dangerous assault landings such as had seized Seria. Helicopters could land troops on any piece of open ground with a diameter a few feet wider than the span of their rotor-blades. At Suez, the Marines had made an assault landing from helicopters and now this new and revolutionary technique would be available in Borneo.

In Brunei, a State War Executive Committee, such as had been set up during the Malayan Emergency, had had its first meeting on the 13th. Under the chairmanship of Sir Denis White, the High Commissioner for Brunei, who had been in England when the revolt broke out, this was attended by the Governor of Sarawak, senior police officers and Brigadier Glennie. While Brigadier Patterson commanded the land forces in the field, Glennie had remained in overall command, answerable directly to the Commander-in-Chief, Far East, with the title of Commander, British Forces, Borneo, or, in military jargon, COMBRITBOR. Walker was now to relieve Glennie, who would return to his desk at Far East Headquarters.

Walker was impressed by Poett's grasp of the current situation and by the steps he had taken to set up the force and its command structure in Borneo once the initial surprise and confusion had been overcome. It seemed likely that Poett had put to good use his experience as Chief of Staff to General Harding in the early years of the Malayan Emergency.

Walker seemed less impressed by Poett's opinion that the campaign should be over, and he would be back in Malaya, within three months. Although the Commander-in-Chief, Land Forces, had visited Brunei and he had not, he remembered the false optimism at the outbreak of the Malayan insurrection and he knew that a guerrilla or terrorist campaign is as easy to prolong as it is difficult to put down. Even if it were confined to the present area, the campaign would, Walker feared, be a long one. He thought in terms of years rather than months.

Walker took the night train to Seremban to collect his uniforms and kit and to say goodbye to Beryl. Next morning he flew back to Singapore and on to Brunei. He took over as COMBRITBOR on 19th December.

Walker had once before been to Borneo and what little he knew of the country and its people had mostly been learned in conversation with his brother-in-law. During his time commanding 99 Brigade and the 17th Division he had been required to think either of internal security in Singapore and Malaya or operations in support of allies to the north. Even so, he felt confident that much of his experience in the Malayan Emergency and his study of insurgency in Indo-China could be applied to Borneo. And he was certain that the close liaison he had always insisted upon between the Services, the police and the civil administration – what he called 'jointmanship' – must be the basis of his thinking.

Impatient to get to grips with the problems ahead, Walker occupied the three hours of the flight to Brunei in jotting down notes for a directive he planned to issue immediately he arrived.

He wrote: 'The ingredients of success shall be five-fold. 1st – Jointmanship. 2nd – Timely and accurate information; i.e., a first-class intelligence machine. 3rd – Speed, mobility and flexibility of Security Forces, particularly Army. 4th – Security of our bases, whatever they may be, wherever they are; whether an airfield, or patrol base, or whatever. 5th – Domination of the jungle.'

As the aircraft began to descend, Walker looked down at what was to be his battlefield. A coastline matted with mangrove swamps and nipah palms and riddled with creeks, inlets and, often, broad, brown rivers twisting through the solid green mat of jungle treetops. As he expected, there were no roads to be seen and only an occasional trail leading from a riverside longhouse into the trees. Inland, hills heaved up under the jungle and beyond stood the peaks of mountains, rock sometimes breaking through the suffocating forests. Then, there was the Brunei river and the capital itself: the gilded dome of the great mosque, the white walls of the Istana and the huddles of stilt-dwellings out in the shallows and on the mudbanks.

On landing, Walker was greeted by Glennie. The immediate task was the briefing of the new Commander. Glennie went over the history of the campaign from its shaky start. Far East Command had been totally unprepared. Not only had the necessary plans and organisation not been ready – Glennie agreed that Patterson should have been allowed to take his brigade headquarters to Brunei and control operations from the start – but none of the Services had been properly briefed. The Army had lacked maps, the Navy knew little about what port facilities there were and the RAF had not known the location of the few existing landing strips. They had had to learn as they fought.

The present situation had crystallised into two parts: a reality and a threat. The reality was the hunt for the survivors of the TNKU. Some of the 'hard core', including, it was believed, Yassin Affendi, had taken to the jungle – probably the maze of waterways in the mangrove and nipah swamps near Limbang – and were being tracked by Gurkhas. Many of the more resolute rebels were trying to make their way to Indonesian Kalimantan and attempts to intercept them were being made by the Army and Harrisson's Force of Kelabits with notable success. The bulk of the rank and file, many of whom had joined the rebellion in the vague belief that they were defending their Sultan against some undefined imperialist attack, had discarded their uniforms and returned to their villages.

The Special Branch of the Brunei Police, which had been so weak, had been reinforced and there was an increasing number of intelligence officers in the field. Their efforts were building up a picture of the rebel order of battle and fugitives were being tracked down in large numbers both in the jungle and in their villages.

Tom Harrisson's knowledge of the interior was the key to stopping the bolt-holes into Kalimantan. He and his Kelabits knew all the trails, for they had used them when fighting the Japanese. A man of strong personality and opinions, Harrisson initially had many arguments with Army officers who, he thought, were trying to fight an unconventional campaign with conventional tactics. He accused them of trying to sweep a jungle area as if it was a forest in Germany, instead of realising that the enemy and themselves could only penetrate it where there were trails, and only he and his tribesmen knew where these ran. But Harrisson and Walker took to each other at once, partly because of the latter's flair for making use of other men's expertise and partly because he liked Harrisson's zest and imagination. He would spend the day up-country, flying between longhouses, collecting intelligence, and be back in Brunei for the evening conference.

The threat was provided by the Chinese and the Indonesians. In Sarawak, it was known that many thousands of Chinese in the towns were members of the Clandestine Communist Organisation and that they were sympathetic both towards the Brunei rebels and to Indonesian opposition to the Federation of Malaysia, the formation of which was due next summer. They had not as yet staged demonstrations, labour disruption, sabotage or other opening moves of a terrorist campaign. Yet they were capable of doing all these things, or shortly would be. They were also capable of providing a power base from which others could operate. While the long-term aims of the Chinese and the Indonesians in Borneo were divergent, their short-term aims – notably the sabotage of Malaysia – were identical and it was clear that they could together form a dangerous, if temporary, alliance.

The Indonesian threat was also vague but gradually coming into clearer focus. The immediate risk was not only that they might harbour fugitive rebels from Brunei in Kalimantan but that they might train and arm further volunteers and foster subversion in all three Borneo territories. Intelligence reports were suggesting that already there might be up to one thousand volunteers, as yet poorly trained and armed, encamped in Kalimantan and available for forays across the border. Behind them were the Indonesian regular forces and these seemed to have been reinforced recently to bring their strength from about six to eight thousand. There were reports and rumours of troop movements towards the border in Kalimantan but, as yet, not one confirmed report of any actual incursion.

At this time, the threat within North Borneo was concentrated around Tawau at the eastern extremity of the coastline where the Royal Navy had been engaged in the romantic but dangerous task of putting down piracy.

Here, Indonesian workers were employed in the timber forests and there was an Indonesian population of about thirty thousand. Although there had been no trouble, the potential was there.

Three weeks after the revolt, a considerable British expeditionary force had been built up in Borneo. In Brunei and East Sarawak, the scene of the rebellion, Patterson's 99 Gurkha Infantry Brigade now consisted of five major infantry units – the Queen's Own Highlanders (less one company), 1st Green Jackets, 40 and 42 Commandos, Royal Marines (less one company) and the 1/2nd Gurkha Rifles. In

West Sarawak, where the CCO threat was worst, there was the head-quarters of 3 Commando Brigade plus one company and a Royal Artillery battery in an infantry role. In the Tawau area of British North Borneo there was one company of the Queen's Own Highlanders.

This force was supported by the Brunei Army, the Sarawak Rang-ers, the constabulary of the three territories and the four thousand men of Harrisson's Force.

The naval and air forces were, by their nature, flexible and vari-able in strength. The Royal Navy's most valuable contribution was undoubtedly the commando ship *Albion*, which could not only operate and maintain a helicopter force but could also act as a fast ferry for light aircraft, vehicles and equipment from Singapore. A few frigates were available for use in ferrying troops, shore bombardment and pro-viding coastal radar cover. For inshore operations, the Navy could only offer coastal minesweepers, small wooden ships capable of only fifteen knots, so unable to reach half the speed of the fast gunboats the Indonesians were known to have, or the power-boats that might be available to insurgents. The reliance on these ill-suited ships illus-trated the Admiralty's folly in scrapping its high-speed Coastal Forces after the Second World War.

The number of aircraft fluctuated but generally there might be five medium-range transports (Beverleys and Hastings), nine short-range transports (Twin and Single Pioneers) and eighteen helicopters (RAF Belvederes and Sycamores and Fleet Air Arm Whirlwinds and Wes-sex) available.

All three Services came under the direct command of Walker as Director of Operations and he himself was answerable directly to the Commander-in-Chief, Far East, so by-passing the three Service Com-manders-in-Chief. Walker was determined to make his own headquarters small, efficient and inter-Service. When he arrived, the central Operations Room was in the High Commissioner's office in Brunei but the three Ser-vices still had their own headquarters as far apart as they had been in Sin-gapore. The Army's was in Brunei, the Navy's at sea and the RAF was on

Labuan island. So, three days after his arrival, the new joint headquarters was set up in a newly-built girls' school, commandeered for the purpose, in Brunei. Walker himself occupied the headmistress's study.

Next day, the Air Task Force Commander moved into HQ, COM-BRITBOR, and the naval liaison officer worked in the Army operations room. Walker made it plain that all the reins must be drawn together in this headquarters and that they came directly into his own hands.

It was not only the three Services' commanders in Singapore who sometimes appeared jealous of Walker's over-riding control of his forces. The Governors of Sarawak and North Borneo were also, officially, the Commanders-in-Chief in their territory and they, too, had to surrender their sovereignty. Soon after his arrival Walker had to make this position clear.

Two advisory bodies had been formed on the lines of those necessary during the Malayan Emergency. One was the Borneo Security Council, consisting of the senior political and police representatives of each territory, the Director of Operations and his senior staff. This body, answerable directly to the Commissioner-General for South-East Asia and the Commander-in-Chief, Far East, was for the formulation of policy.

The other was the Borneo Operations Committee, which under the chairmanship of Walker himself, would discuss and work towards operational decisions.

Three days after his arrival, Walker attended the first meeting of the Security Council, which was attended by the Governors – and Commanders-in-Chief – of Sarawak and North Borneo: Sir Alexander Waddell and Sir William Goode. In outlining his immediate plans, Walker announced that he planned to concentrate his Marines in Brunei and North Borneo and to replace them in West Sarawak with the headquarters and one squadron of the Queen's Own Irish Hussars. Here were some of the few roads in Sarawak, notably those running west from Kuching to Bau and east to Serian and to the frontier police station at Tebedu. These roads ran almost parallel with the Indonesian frontier, like a barrier to the south of Kuching, and they would be intensively patrolled by the Hussars' armoured cars.

Waddell objected. He wanted infantry not armour, he said, and, anyway, he did not want armoured cars churning up his new roads which were so expensive to maintain. Walker replied that the Hussars could fight on their feet as well as in their armoured cars, which should be regarded not only as fast personnel-carriers but as mobile wireless sets.

Waddell remained adamant. He wanted infantry, not armoured cars to defend Kuching. Walker instantly brought the conflict to a head. Turning to Sir Denis White, the High Commissioner for Brunei, he said, 'Chairman, I wish to go on record as having been told that my official military advice is unacceptable to the Governor of Sarawak.'

White agreed that this seemed to be the case but Goode intervened and with diplomatic skill smoothed over the differences and it was agreed that the Hussars should go to West Sarawak in the role of mobile infantry. Walker was amused some months later, when the Hussars had proved their worth, to be told by Waddell that one of the wisest decisions he had made was to ask Walker for a squadron of armoured cars. But by this time his relations with Waddell were as excellent as they had also become with Goode and White.

But at this meeting it seemed to Walker and his staff that the political members of the Council saw the crisis in terms of a revolt in Brunei which was being crushed and had not fully appreciated the wider implications and dangers. To him, the Brunei revolt seemed not an isolated explosion but a fuse which might ignite a conflagration such as was already beginning to engulf Indo-China.

From the first, Walker was determined to spend as much time in the field as in his office, even if this meant working most of the night. He had liked Harrisson for his eagerness to work in this way and, during this first week, the two men flew together the full length of the Indonesian frontier, nearly one thousand miles from the sandy beaches of West Sarawak to the mangrove swamps of eastern North Borneo. Their light aircraft landed where it could and Harrisson introduced Walker to some of the tribal leaders and briefed him on the background and the characteristics of each tribe. Between stops, the two men looked down on to jungle-covered hills and mountains while Harrisson pointed out the routes across the frontier which he had already marked on Walker's map.

If the Indonesian threat did develop, the problem would be, on the face of it, almost insuperable. With enemy raiding parties able to strike across a thousand miles of frontier from secure bases how could the defenders hope to stop them, defend every potential crossing point? But, flying along the frontier with Harrisson, the task did seem possible. Good intelligence was the answer. If warning could be given the moment the Indonesians crossed the frontier, then they could be trapped in the jungle they had expected to cover their advance. The tribes seemed friendly and ready to help. But their information would have to be collected and assessed carefully because there were not enough troops or helicopters to rush ambush parties into the deep jungle at every report of an incursion.

This was clearly a job for the Special Air Service Regiment and other infantry – British or Gurkha – who had been trained in long-range penetration and intelligence-gathering. A thin but sensitive screen was needed along the full length of the frontier. Once it was known where the enemy had crossed and what their objective seemed to be, they could be dealt with by ordinary, jungle-trained infantry. To Walker, this seemed primarily a task for his Gurkhas.

Although Walker's thoughts had been concentrated on the conclusion of one campaign and the possible opening of another, his visit to Nepal and the fate of the British Army's Gurkhas were never far from them. The events of the past month had convinced him that he was right to oppose the plans of the Government and Army Council by every available means, even if this laid him open to charges of disloyalty and of abusing Hull's confidence. Without the Gurkhas, the Brunei revolt could hardly have been contained.

Now, if he had to face an Asian enemy to whom the jungle was home, along a front of one thousand miles, the Gurkhas would be more important than ever. Indeed, without them, Walker was convinced, Britain might as well cut her losses and abandon all thoughts of exerting influence on the development of South-East Asia.

So, late at night, after his headquarters staff had finished the sixteen-hour day that they had come to regard as normal, Walker wrote a detailed directive, describing his visit to Katmandu, his audience with the King of Nepal and his meeting with the American ambassador. This he planned to circulate as widely as possible within the Brigade of Gurkhas to encourage them with the news that they still had powerful friends. For, in the short term, Walker was concerned that the Gurkhas were being committed to action in a state of uncertainty, not knowing whether the Government that was sending them to battle would, once they had won it, discard them.

Before Walker could complete and distribute this directive, the crisis that smouldered in Whitehall, Katmandu and Gurkha battalions all over the Far East, flared up, ignited from a totally unexpected quarter. In London, on the morning of 3rd January, General Hull read with rising anger a long report in the *Daily Telegraph* from its Military Correspondent, Brigadier W. F. K. Thompson, reporting from New Delhi.

Under the headlines NEPAL CONCERN AT FATE OF GURKHA BRIGADE and LARGE CUT WILL IMPAIR GOODWILL TO BRITAIN, Hull read what he could only assume had been inspired by Walter Walker.

'A spokesman for the Nepalese Government in Katmandu,' he read, 'yesterday told me of the concern in Nepal at the apparent British intention to reduce the strength of the Brigade of Gurkhas. The concern is not only at reductions but also at the fact that so far no approaches have been made to the Nepalese Government.

'Four years ago, Field Marshal Festing (then CIGS) visited Katmandu to put the British suggestion that the five-year renewal of contract for Gurkha recruitment be extended to ten years to give greater stability. The Nepalese Government accepted the suggestion.

'The apparent decision of the British Government to reduce the strength of the Gurkhas without consultation appears to the Nepalese to savour of sharp practice. It was put to me that large reductions would lose Britain long-standing goodwill. It would also inhibit expansion if later required.'

Brigadier Thompson went on to point out the potential risk in the return to Nepal of discontented, discharged soldiers. He claimed that the economic reliance of Nepal upon her soldiers with the British Army – the £850,000 Britain spent on the Dharan depot the previous year had been almost the only free international exchange available to Nepal – was not fully understood.

Although India employed more than forty thousand Gurkha soldiers, the link with Britain gave Nepal added stability. 'Nepal is in an uncomfortable position between India and China,' wrote Thompson. 'She looks to Britain as an old and experienced friend for support and advice. But there is reason to doubt whether she gets it in the measure deserved....'

Brigadier Thompson's argument tallied so closely with that used by Walker that Hull was convinced that he had inspired the report. Already the effects of Walker's audience with the King and his alliance with the American ambassador were throwing out seismic shockwaves that had been recorded in the Foreign Office, the Commonwealth Office and the Ministry of Defence. Walker, it seemed, had not only broken his word to Hull as a gentleman but had disobeyed an order as an officer.

Brigadier Thompson's report echoed Walker's views so exactly that he might well have written it himself. In fact, he had not at that time met Thompson, had had no hand in inspiring this or any other newspaper report and his first knowledge of this one was when, about a week later, he received an airmail copy of the *Daily Telegraph*.

Now Walker had completed his own report.

This went over some of the ground covered by the assessment he had circulated in March but now, he wrote, the Government intended to cut the Brigade of Gurkhas by more than half, leaving only four infantry battalions – about four thousand men – and there were signs that this might lead to the extinction of the British Army's Gurkhas. Walker then reported in detail his visit to Katmandu, describing his audience with the King and Nepalese concern at the news he had to give them. But, he added, he had made a powerful ally in the American ambassador.

The directive was intended both to warn and to encourage. It warned of the British Government's duplicity in preparing to cut the Brigade savagely and secretly, so presenting it and the Government of Nepal with a *fait accompli*. It encouraged by calling for resistance and maintaining that all was not yet lost and, with the help of friends in Katmandu and Washington, the Brigade might yet be saved.

Walker sent copies of the directive to his brigade commanders, the commanding officers of all major Gurkha units and to senior members of the Brigade staff. He also sent a copy to Major-General Lewis Pugh, who was Colonel of the 2nd Gurkhas and had been Chief of Staff, Far East Land Forces. At this time Pugh was Chairman of the Council of Colonels of the Brigade of Gurkhas and had been rallying support for the Gurkha cause among Members of Parliament. Walker did not, however, send copies to the three field marshals who might have been expected to help. Slim, the former Colonel of the 7th Gurkhas; Harding, former Colonel of the 6th; Templer, Colonel of the 7th; could all wield influence in Whitehall but had not done so. It seemed probable that they were waiting for what they considered the appropriate moment to make their views felt, perhaps failing to appreciate that, when they did, it might be too late.

Once the directive had been circulated, Walker realised that he had done all he could for the time being and concentrated on the redeployment of his force.

Although the hunt for fugitive rebels continued successfully there were fresh indications that the Indonesians might try to 'stir the dying embers of the revolt', as Walker put it. This might be by the infiltration of agents into Brunei, it was thought, or even by raids across the frontier. By now, two new threats had developed. There were ominous signs of Indonesian military activity at either extremity of the frontier with Kalimantan.

Both Tawau in the east and Kuching in the west were vulnerable to a *coup de main*. Units up-country were therefore strengthened and contingency plans made. At both Tawau and Kuching, vehicles and ammunition were stockpiled and the RAF practised flying an infantry company group to either place at six hours' notice.

During the coming month there was to be a major reshuffle of units, including the relief of the Queen's Own Highlanders and 1/2nd Gurkhas by the King's Own Yorkshire Light Infantry and 1/7th Gurkhas. Redeployment was about half complete when the weather broke. It rained as it had not rained in living memory. In one week an average sixteen inches of rain fell over Brunei and the Fourth and Fifth Divisions of Sarawak. Borneo is accustomed to monsoon storms and it was usual for a sudden downpour to swell the rivers so fast that in the hills their level might rise twenty or thirty feet in a few hours. But this time the rivers rose over their banks, inundating vast areas. The rain virtually stopped the war and, on 17th January, Walker ordered that flood relief take priority over everything, not only redeployment but even operations against the rebels. He made medical supplies and food freely available, telling his staff that he took full responsibility and would account for the cost to Singapore himself.

The three Services instantly switched from taking lives to saving them. On the Limbang river, which had risen fifty feet, the Royal Navy and Gurkhas manned rescue boats. The RAF parachuted food to towns cut off by the floods and Fleet Air Arm helicopters flew in rice and flew out refugees. In most towns power and water supplies had failed and the Services provided generators and purification plant.

Relief and rescue work was carried out with the urgency and intensity of a warlike operation and there were many acts of gallantry, often unrecorded.

By the end of the third week in January, the floods had subsided and life began returning to normal. But there had been one change that would persist, a change that Walker instantly recognised and used to his advantage. Virtually the whole population of the stricken area had either seen the Services working at relief and rescue or had heard about their efforts. Now, where there had been indifference there was friendliness and where there had been friendliness there were now allies. Walker was determined to keep this friendship and increase it. A major military aim in Borneo must be to win the hearts and minds of the population.

So began what came to be called the 'Hearts and Minds' campaign. This was something new. In Malaya this had not been done until the arrival of Templer. The Briggs Plan for the resettlement of villagers had involved at least inconvenience and at worst distress. In this campaign, the Services must give a high priority to civil aid: medical treatment, help with small engineering projects, air transport whenever possible and, above all, courtesy and friendliness. Here Harrisson's advice was invaluable. He had some sharp exchanges with Army officers whom he accused of needlessly upsetting the tribes up-country and, on one occasion, he berated a British unit for burning down a longhouse to enlarge their field of fire.

The value of Bornean hearts and minds was to be demonstrated within a week of the end of the floods. From the first, Walker had realised that the only hope of defending a thousand miles of frontier against possible Indonesian incursions was to watch all potential crossing-points. For this purpose he was now sent a squadron – or a company – of the 22nd Special Air Service and, in support, the Gurkha Independent Parachute Company.

The SAS had been founded during the Second World War as a specialist force trained in deep-penetration raids behind enemy lines and sabotage. They were all parachutists but they had also ranged widely across enemy lines of communications in small, heavily-armed, mobile patrols. In the face of opposition from those who thought that the SAS skimmed the cream off the infantry regiments, they survived in peacetime and proved their worth in Malaya where they lived among the aboriginal tribes in the deep jungle and became as adept at living rough as the guerrillas they hunted. They had also been operating in the mountains of South Arabia in support of the Sultan of Oman but, since 1959, the SAS had concentrated on training for large-scale war, particularly in Europe.

The arrival of the squadron coincided with a new policy which was being developed by the regiment's commanding officer, Lieutenant-Colonel John Woodhouse. In 1962, half the SAS flew to Fort Bragg, North Carolina, for training with the United States Army's Special Forces ('The Green Berets'), who were now being committed to the fighting in South Vietnam. As a result, the SAS were learning new skills, notably in medicine and languages. This would not only enable them to make contact with primitive peoples in remote places and live among them but also to repay their friendship with practical help.

Woodhouse and Walker liked each other immediately. Some senior officers regarded the SAS as an adjunct to the Parachute Regiment or as a bunch of individualists who got in the way of real soldiering. Walker, on the other hand, first wanted to know what the SAS could do and how Woodhouse would himself react to the threats in Borneo. The troopers liked him as much as did their colonel because he did not preach and was eager to hear their views. The SAS also liked Tom Harrisson with whom they formed so close a relationship that, when he returned to Kuching early in 1963, they set up their headquarters in his garden.

Walker planned to use the SAS as a border screen in four-man patrols equipped with lightweight wireless transmitters. Studying the points to be watched it was clear that, with twenty-odd patrols needed, the squadron would be overstretched. At first patrols were cut to three men – at least one speaking Malay and another trained in basic medical skills – and the Gurkha Independent Parachute Company was to be spread out in similar small patrols through the interior between the frontier and Brunei and Kuching. The SAS patrols were to be self-sufficient for four weeks and then would be supplied by air so that they could remain on station for up to four months. Among the parachuted supplies would be sugar, soap, salt and kerosene, all valuables in the interior, as gifts to reward the friendship of the tribes.

At first some of the SAS men were sent up-country in civilian clothes; an unexpectedly difficult task because the shirts and jeans on sale in Borneo shops were cut for diminutive Chinese figures not hulking British troopers. But disguise was quickly discarded as unnecessary and the patrols wore the usual jungle-green denims and soft-brimmed hats with a yellow band for identification. In jungle fighting, identification must be instant and reaction so quick that, as there might be time for only one shot at an enemy, the 'point man' of each patrol was armed with a large-bore shot-gun with a wide spread of heavy shot.

One of many problems facing the SAS screen was the paucity and inaccuracy of maps. To the north of the frontier, mountains might be misplaced by twenty miles and rivers as wide as the Thames at Richmond not marked at all. On the ground, the frontier was unmarked and two available maps had different versions of where it ran. Of Kalimantan there were no detailed maps at all and, as the first SAS men looked through binoculars into Indonesian Borneo, the country beyond was as strange and unknown as any that has faced an explorer.

The success of the Gurkhas in Brunei and the zest with which they were again taking to the jungle though the Borneo territories led many Gurkha officers to believe that this would now finally remove any danger of serious manpower reductions for years to come. But this was not the case. While, officially, the Army Council spoke only in terms of a possible reduction to ten thousand, the number agreed with Nepal by treaty, their real intentions now seemed worse than even Walker had feared. Throughout the Brunei revolt, Walker had continued to receive privately details of the Army Council's long-term thinking and planning.

There was no officer on the General Staff able to stand up to Hull, Mr John Profumo, the Secretary of State for War, and the rest of the Council, who seemed committed to this course.

Walker believed that General Poett, in Singapore, supported Hull's view. As soon as Walker had been informed of the intention of reducing to ten thousand he had circulated two signals. One, for the benefit of the Gurkhas, was breaking the news as gently as he could and asking them to accept it with their customary loyalty. The other was to the Chief of Staff, Far East Land Forces, stressing the damages and complaining that the affair had been 'handled disgracefully' by the War Office. Poett had read both signals.

But early in 1963 Walker's information was that the Brunei revolt had in no way changed War Office policy.

Then, soon afterwards, the first storm warning arrived. A senior staff officer of the Brigade flew to Brunei from Singapore on a routine visit, and told Walker that Poett had been making enquiries about a certain paper he had circulated within the Brigade. This, he thought, was probably Walker's directive following his visit to Nepal because that, which had been regarded as an internal affair for the Brigade, had not been sent to the Commander-in-Chief, Far East Land Forces. From what he gathered, it was believed that Poett had got hold of a copy of the directive and was anxious to learn about its origins and circulation. This had, of course, been strictly limited to within the Brigade and a copy had been sent to Walker's old friend Major-General Lewis Pugh, a former Chief of Staff, Far East Land Forces, who was now retired in Wales.

Soon after, in March, Walker received a non-committal signal from the War Office summoning him to London. The reference number on the signal bore the prefix MS, the initials of the Military Secretary, which showed that it had come from the office of the Chief of the Imperial General Staff. Could this mean that Hull, too, had either seen or heard about Walker's directive and that this might be the reason for his recall?

So regretfully, but with confidence, Walker left the war in Borneo in the capable hands of Glennie, flew to Singapore and was given priority on a flight to London. With difficulty he thrust the campaign to the back of his mind while he considered what might have happened. The only copy of the directive which he knew to be in Britain had been sent to General Pugh and he decided to contact him as soon as he arrived and before reporting to the War Office.

On arrival, Walker took a train to North Wales and presented himself to Pugh. He came straight to the point. Had Pugh shown his directive to anybody in, or connected with, the Army? Pugh said that he had. As an old friend of the Gurkhas, he had been appalled by the threat to their existence and felt it his duty to urgently recruit support for Walker's stand. He had therefore shown the directive to Field-Marshal Sir Gerald Templer, who preceded Hull as CIGS and, as Colonel of the 7th Gurkhas, could provide powerful support. But Pugh had mis-read Templer. As a former CIGS, his instinctive first loyalty was to his successor and, as a Field-Marshal he was still a serving officer whose loyalty lay with the Army Council. Moreover, Templer was a stickler for military discipline and correct procedure.

After reading the paper, Templer had told Pugh that he considered Walker to have been foolish and incorrect in writing and circulating it. He had, however, decided to show it to another Field-Marshal, who had also been CIGS and also Colonel of the 6th Gurkhas: Field-Marshal Lord Harding.

What Pugh did not know was that Harding had been as upset as Templer and had passed the directive to the doyen of Field-Marshals, Chiefs of the Imperial General Staff and Gurkha officers, Field-Marshal Lord Slim. At this time, Slim was the arbiter of all things military – a brilliant fighting general, a great leader and an honourable man – and his views were accepted without question by the majority of his contemporaries in the Army. Slim had Walker's directive photo-copied and, after annotating its margins with such comments as 'disloyal' and 'disgraceful', had forwarded it to Hull. What was particularly ominous was that his covering letter did not begin with his customary 'My dear Dick' but with 'Dear CIGS'.

Walker appreciated that Pugh had acted with the best intentions but he made it clear that he had sent him the directive in strict confidence and had not expected him to show it to anybody without first asking his permission. But he recognised that he, too, could be charged with similar conduct, for clearly Hull must now regard him in the same way. He would soon find out.

Returning to London, Walker found a message waiting for him at the United Service Club in Pall Mall from Lieutenant-General Sir William Pike, the Vice-Chief of the Imperial General Staff. This invited him to call next day and confirmed that Hull wanted to see him on the day after. Pike was an old friend; he had begun his career in the Indian Artillery and had been Chief of Staff, Far East Land Forces, when Walker had commanded 99 Brigade. The two had always trusted one another.

Next day, Walker set off down Whitehall, his bowler hat and umbrella slightly incongruous against his sallow tropical tan. At the War Office, Pike greeted him warmly with talk about his success in Operation Tiger three years before and how highly he was regarded as a field commander. Then he said what Walker had expected.

Hull had read the directive and was very angry indeed. Walker had been recalled from Borneo to answer for his action. Hull regarded this as gross disloyalty. Profumo was outraged at his alliance with the American ambassador.

'I know why you did it, Walter,' said Pike. 'I can read you like a book. You are a fighter and you are fighting for your Gurkhas with every weapon you can find. But you have made your point and you cannot now fight the CIGS. I ask you to accept defeat and take it on the chin.'

Walker replied that he saw no reason to stop fighting for his Gurkhas now. If he did not fight, who would? They had no Member of Parliament to take up their cause. As Major-General, Brigade of Gurkhas, it was his duty to fight.

Pike repeated that it was impossible to fight the CIGS. 'If you do, Walter, he will sack you. He will have no alternative. And then what would happen to your Gurkhas? And what would happen to the campaign in Borneo? You have achieved your aim. Now you can afford to say you are sorry.'

Reluctantly, Walker agreed that he would.

As he left the War Office deep in thought, Walker ran into an old friend, one he remembered as being up-to-date in Army gossip, the more scandalous the better. Had Walker heard the latest? Did he know Jack Profumo and what he had been up to? Well, he and a Russian naval attaché had been sharing the same girl. What about that, now?

Walker was too absorbed by his own dilemma to pay much attention and, in any case, this was the sort of chatter which never amused him. There were far more important subjects to discuss.

That night he spent with his sister, Beatrice – 'Bee' – and her husband, Roy Hudson, who had served on Montgomery's staff during the war, and confided in them. What should he do? Hudson agreed with what Pike had advised. Walker's stand and events in South-East Asia had surely saved the Gurkhas. Now he could afford to apologise.

Next day Walker returned to the War Office.

The time for the confrontation had arrived. Along the corridor, Walker entered Hull's outer office and his hat and umbrella were taken by the Duke of Kent, who was a junior member of the staff, and sat down to wait. He waited for forty minutes. Then he was told that the CIGS would see him now.

Hull was standing with his back to the door, looking out of the window. He turned, made no move to shake hands and said without a smile, 'Sit down, Walter.'

Hull said that he had been amazed by a letter he had had from Field-Marshal Slim and by Walker's directive which had been attached. Slim had been so outraged that he considered Walker as unfit to continue serving. He, too, had been astonished that Walker should betray his confidence. Profumo had been furious at the involvement of the Americans.

'Why did you do it, Walter?' asked Hull. 'I don't understand you. I never expected such behaviour from you. You seem to have developed a split personality.'

When he had first heard of the directive, Hull continued, he had been on the point of leaving for a tour of the Far East. At first he had considered summoning Walker to Singapore and relieving him of his command there and then. But he had discussed the matter with Poett and they had decided that it would be best to summon him to London and bring the matter to a head there.

'Why, Walter, did you adopt these miserable tactics?' asked Hull.

Walker knew that his job as Director of Operations, even his career, was in the balance. He did not know that Hull had a far more serious consequence in mind: a court-martial. For he regarded the issue as one not only of betraying a confidence but disobeying an order.

Some of his colleagues regarded Hull as an old-fashioned officer, conventional in his social attitudes. Indeed, one brigadier who watched the affair from close quarters thought that Hull's anger was really directed at a breach of accepted social conduct – conduct unbecoming to an officer and a gentleman. Walker, on the other hand, had behaved much as senior American officers customarily behaved: rallying a powerful lobby to produce results which conventional methods could not.

Much in Hull's mind was his own experience nearly twenty years before. Then, as a young but highly-placed officer on the General Staff in London he had worked on plans for the disbandment of the 1st Armoured Division, which had been through heavy fighting on the

Continent. A matter of weeks before the date set for this, he was told that he was to take command of the division but, to maintain morale though the battles immediately ahead, he must keep to himself the knowledge that their days were numbered. This had been a great strain because his brigadiers and colonels had trusted him and he could not confide in them. But he had had to do it and had done it. If he could do this in the midst of war, then surely Walker could have done the same during a limited internal security operation.

Walker longed to hit back. He wanted to say that, while Hull might be justified in accusing him of breaking a confidence, he and the Army Council were doing just that to the Gurkhas, to the King of Nepal and to himself. But, at the time, he could not admit that he knew much of what the Army Council had been discussing in private for fear of betraying his sources. He therefore, as Pike had advised, 'took it on the chin'.

Then Walker apologised. He said that he was very sorry to have resorted to such measures and that he fully understood Hull's feelings about it.

Hull replied, 'Had you not apologised I would have had to sack you. I did not want to do that for the sake of the Gurkhas and the effect the news might have on them at this time.'

But, he added, Walker would no longer advise the Army Council as Major-General, Brigade of Gurkhas. He would not be relieved of that appointment but, for the rest of his term, this responsibility would go to Major-General James Robertson, who was being recalled from retirement to become Gurkha Liaison Officer at the War Office.

The interview was at an end. Before leaving the War Office, Walker called at Pike's office to let him know that he had taken his advice. There he was introduced to a naval officer, a square-shouldered man with a strong, lined face which broke into a wide, friendly smile. This was Admiral Sir Varyl Begg. They had not met before but each knew exactly who the other was. 'I look forward to seeing you next month,' said Begg. He was about to relieve Admiral Luce as Commander-in-Chief, Far East.

Walker liked him at once. To return to Borneo as Begg's Director of Operations was another good reason for trying to make peace with Hull.

That evening, his brother-in-law said that it would be prudent to put a final end to the conflict by writing a letter of apology to Hull before leaving London, and this he did. One more task remained: to order all those who had received his directive to destroy them.

It seemed to Walker that the crisis had passed. He had had to apologise against his inclination and accept the resulting loss of face. But if this had saved the Gurkhas – as he was convinced it had – then he was well satisfied.

A few days later, he was back at his headquarters in Brunei. Ostensibly, the only change was that now General Robertson was regarded by the War Office as the new Major-General, Brigade of Gurkhas in all but name. Walker did not accept this as he made plain when, soon after, Robertson visited him in Borneo. They were old friends but the position was made clear as soon as Robertson began discussing Gurkha affairs.

'Remember that I am still Major-General, Brigade of Gurkhas,' Walker told him. 'You have been appointed adviser to the Army Council on Gurkha affairs. But I am wholly responsible for the Brigade of Gurkhas itself. I run all the Gurkhas' internal affairs. If you put your finger in my business, I'll cut it off.'

The Brunei campaign was nearly over. The final mopping-up operations were going well and most came to a sharp climax with the killing or capture of fugitive rebels. Then, in May, a patrol of the 2/7th Gurkhas, guided by an informer, waded through a mangrove swamp and flushed a gang of rebels from their bivouac, driving them towards other Gurkhas lying in ambush. A few bursts of fire and it was over. Ten rebels were killed or captured. They were all that had been left of the TNKU headquarters. One of the rebel wounded was identified as Yassin Affendi. At that moment the Brunei rebellion was officially over.

Even before that moment, Walker had assessed the campaign, passing on the lessons that he believed should now have been learned. One of the foremost of these, he maintained, was 'the need for the retention of HQ 17 Gurkha Division and HQ 99 Gurkha Infantry Brigade Group as field force HQs with training and operational responsibilities, particularly in view of the limitations imposed on 28 Commonwealth Brigade, has been amply demonstrated'.

Only such headquarters, based in the Far East and trained for operations in that theatre were of value. It would be useless to maintain a quick-reaction force in Britain because it would arrive on the scene of action too late, unacclimatised and unprepared. His theory was unpopular in Whitehall where the idea of a home-based 'fire brigade' strategic reserve had become popular as an alternative to maintaining a strong military presence in the Far East.

Indonesia was now the immediate threat, Walker believed, and President Sukarno would probably 'use more than words' in an attempt to prevent the formation of Malaysia that summer. The Russian Defence Minister, Marshal Malinovsky, had recently visited Jakarta, he pointed out, and he was unlikely to have been trying to bring peace to South-East Asia.

In Singapore, meanwhile, Headquarters, Far East, shared Walker's forebodings in general but not in detail. The trouble in Borneo appeared to have been put down decisively and the tension on the Indonesian border and disquieting reports from within Kalimantan seemed to be mostly sabre-rattling propaganda. Indeed, Walker was accused of over-playing his hand by reporting that Indonesian forces there had been strengthened by two thousand men. Walker had, in fact, said that these reinforcements had moved into the frontier area and not that they had arrived in Kalimantan to join the eight thousand Regular fighting men already there.

Yet GHQ had other threats on their minds. The Indonesians might try to make trouble in Singapore and Malaysia, perhaps landing raiding parties and saboteurs to join Chinese dissidents in attempting to disrupt the coming political union. The guerrilla war in Vietnam was intensifying and might spread through Laos and Cambodia into

Thailand. The threat of trouble in Borneo was seen by Headquarters as one of several and by no means the most dangerous.

But Walker's insistence on the importance of using his own divisional headquarters to meet any sudden outbreak of trouble was now accepted. Yet, ironically, his success in making this point reacted against his own plans. The place for him and his headquarters was now at the centre and Borneo seemed quiet enough to be left to Brigadier Patterson. So, early in April, Walker was ordered to return with his headquarters staff to Seremban and hold itself in readiness for another emergency.

Preparations for the return were complete and Walker was about to make his farewells when on Good Friday, 12th April, an urgent signal reached his headquarters in Brunei. A duty officer telephoned Walker to tell him that a police station at a village called Tebedu in Sarawak had been stormed and captured at dawn that morning.

The attackers had come from Kalimantan. They had been Indonesians.

This was what Walker had expected. The rebellion was over but the war had begun.

6

The Secret War

In itself the raid on Tebedu was a minor affair but it was to have far-reaching consequences. The police station, three miles north of the Indonesian frontier and due south of Kuching, had fallen to some thirty raiders who had crawled along a culvert beneath the surrounding wire fence and taken it by surprise. In the brief burst of firing, two policemen were killed.

When the news reached him in the early hours of Good Friday, Walker did not know the identity of the attackers. There was known to be a strong cell of the Clandestine Communist Organisation among the Chinese in Tebedu village and it was possible that they had launched the attack to seize police firearms. On the other hand, a well-planned commando raid such as this had been could also have been carried out by Indonesian regular soldiers from Kalimantan. This was, in fact, what had happened but Walker did not receive proof until later. Meanwhile he assumed the incident to be as serious as it could be and acted quickly.

First, reinforcements must go to the First Division which was held by only a company of 42 Commando and the armoured car squadron of the Queen's Own Irish Hussars. The stand-by company in the Brunei area was out in the jungle with the rest of the King's Own Yorkshire Light Infantry and had to be collected from jungle trails by a light aircraft circling overhead, calling to them by loud-hailer. RAF aircraft were ready to fly them to Kuching but, once there, they would need helicopters.

At Brunei airfield the six Whirlwinds of 846 Squadron, Fleet Air Arm, were undergoing maintenance checks and one helicopter had had its engine removed for an overhaul. Walker asked how soon they could leave for Kuching and was told a week. He ordered the squadron to leave next day.

That same evening the infantry flew out, followed next morning by the helicopters which, before landing, flew over Kuching in formation. Meanwhile, 3 Commando Brigade, which had recently returned from Sarawak to Malaya was ordered to return. On Easter Sunday, the first company of 40 Commando flew to Kuching, their headquarters arriving next day and more reinforcements, including the 2/10th Gurkhas following in the commando ship *Albion*.

A mobile striking force of one commando troop and two troops of armoured cars was quickly formed and rushed to Tebedu. The police station had been hastily evacuated by the raiders and the commandos occupied and fortified it. The incident was over but a new and dangerous situation had begun to develop.

Walker was grateful for the Tebedu incident because it proved to his superiors in Singapore that his apprehensions were well-founded.

Two days after the raid, a team from GHQ, Far East, led by General Wyldbore-Smith, the Chief of Staff, had flown to Kuching and confirmed Walker's fears. Now action could be taken not only against immediately obvious military threats but against those which some had regarded as imaginary. The most serious of these was the CCO in the First and Second Divisions of Sarawak.

In Walker's view, the Tebedu incident could not be seen in isolation. It had, he learned, been carried out by trained Indonesian soldiers and was surely part of a strategy. This, he believed, was the active support of dissidents within Sarawak. There was already unrest, there could now be insurrection.

The raid had coincided with a report on the CCO from the recently-strengthened Special Branch of the Sarawak Police. This showed that the communist underground was far stronger and more widely-spread than had been thought. Now Walker could command the full support for the preventive moves he had been urging. The first of these was the strengthening of police station defences with barbed wire, sandbagged strongpoints and extra sentries. Although Director of Operations, he lacked the comprehensive powers which Templer had enjoyed in Malaya. He had no political power over the governments of Sarawak, Brunei and North Borneo and no overall control of the police.

Now the Sarawak Emergency Committee agreed that their police and Walker's soldiers should combine in a major operation against the CCO. The breaking down of its organisational structure would clearly be a long task but immediate, practical steps could be taken to reduce its strength in weaponry. The Special Branch considered that, while CCO cells and communications between them and into Kalimantan were already working efficiently, they lacked the arms needed for a general uprising. However one weapon of which there was no shortage was the twelve-bore shot-gun for which more than eight thousand five hundred licences had been issued by the Sarawak Police.

On 19th April, all licenses were revoked and all firearms were, it was decreed, to be surrendered. At the same time, the infantry which had been rushed to Sarawak carried out a series of raids and searches for hidden arms. By the end of the first week nearly eight thousand shot-guns had been collected.

These raids confirmed that, while the CCO was not yet equipped for a guerrilla campaign, it was well organised and capable, at least, of carrying out sabotage and selective assassination. This possibility and outbreaks of violence along the frontier convinced some that this would become a repetition of the counter-terrorist campaign in Malaya. Walker agreed that, at that time, military and police operations could usefully be modelled on the Malayan experience. But there was a fundamental difference. The Malayan Emergency had been an internal disorder. Here the threat was both internal and external and the combination of the two could develop into a situation far more serious that anything faced during the twelve years' fight against the Malayan Chinese guerrillas.

Walker saw his task neither in terms of internal security nor of limited war but in the orchestration of means to face both or either. To do this it was vital that he should be the unchallenged conductor. Templer had been this in Malaya but now Walker lacked complete control over the police in Borneo and the likelihood of his achieving this seemed remote. When Sarawak and North Borneo joined the Federation of Malaysia in the late summer the position would be complicated. Brunei would remain an independent British protectorate but the Governors and Emergency Councils in the other two states would lose their sovereignty to Malaysia and a new command structure in Kuala Lumpur. The police would certainly remain under the control of their Inspector-General, Claude Fenner, who seemed to have no intention of relaxing it. Although Fenner and Walker were friends they differed in their views of the Borneo troubles. In Borneo it sometimes seemed that Fenner thought they were fighting the Malayan Emergency over again.

Where the command structure was untidy, as here, it was obviously essential that Walker should have trust in those who shared some of the responsibilities of containing the dangers. Sadly, this was not always the case.

In attempting to assess attitudes and personal relationships at this time it is important to recognise the existence of Walker's circle. This was no clique nor was it recognised at the time as a cohesive body of opinion. It was simply those officers – in all three Services – who shared confidence in the new unified command system in the Far East and in Walker's efforts to use land, sea, air and police forces as a single weapon. So, while Walker's circle welcomed unified command and the appointment of a Director of Borneo Operations, answerable directly to the new Commander-in-Chief, and by-passing the Service Commanders-in-Chief, they felt that much was still to be learned at the higher levels of command.

After the battle. Soldiers of the Royal Leicestershire Regiment
resting after the action in which they stormed an Indonesian camp,
killing seven of the enemy in January 1964

The war-winning helicopter. A Wessex of 845 Squadron, Fleet Air Arm, lands at a jungle base to lift a patrol into their battle positions

The unified structure had not been welcomed by the majority of senior officers. Traditionally, in peacetime, the three Services had competed for larger shares of the defence budget, often in an open, friendly way, as did the nationalised electricity and gas industries in Britain when each used Government money to advertise its superiority. Inevitably, when cuts in defence expenditure became necessary, cheerful rivalry gave way to ruthless competition. In this, each Service developed its own creeds.

While the Army was required to maintain a powerful corps in Germany to support the North Atlantic Treaty Organisation and also maintain a strong presence in the Middle East and Far East, it was comparatively confident of its future. The NATO role assured the continuing need for armour and artillery and that in the East for infantry. They, too, would have to face cuts and these would be hurtful but – with the important exception of the threat to the existence of the Brigade of Gurkhas – these would not threaten the Army as a body or as a profession.

Fundamental rivalry had developed between the Royal Navy and the Royal Air Force. Air power had destroyed the monopoly of sea power after a long confrontation which reached its climax in the sinking of the British capital ships, *Prince of Wales* and *Repulse*, by Japanese aircraft off Malaya in 1941. But, conversely, the doctrine of air power had failed in its hopes of inflicting decisive defeat upon Germany and, in attempting to do so, had so weakened the sea and maritime air effort that Britain was brought near to defeat by submarine blockade.

Since 1945, the two Services, each dependent for its viability upon technical developments and their cost, had become deeply apprehensive. The Navy proclaimed that the aircraft carrier and its specialised maritime aircraft had become the new capital ship, the prime instrument of sea power. The RAF claimed that nuclear bombs had given air power the invincibility it had lacked before the demonstration atomic bombs were dropped on Hiroshima and Nagasaki in

1945. These two systems were so expensive that it seemed likely that one would have to go, or that both would have to be cut to a feeble minimum. In 1957, the Navy had convinced the Government that carriers remained essential, while the RAF continued its campaign to discredit the concept.

Looking to the not-too-distant future, it was clear that both carriers and the V-bombers of the RAF, which had the deterrent role of carrying strategic nuclear weapons, would become obsolete. What then? The RAF hoped to prolong the life of the airborne deterrent by mounting long-range, strategic missiles – such as the American-developed Skybolt – on its aircraft. The Navy was showing a lively interest in a revolutionary concept: the launching of long-range Polaris missiles from submerged submarines. The nation could not afford both, so competition became increasingly intense.

Thus, in the Far East as in Whitehall, the protagonists preached with mounting fervour the rival doctrines of Sea Power and Air Power.

After his arrival in Borneo, Walker received his first lesson in sea power from the Royal Navy. An early visitor to Brunei was Admiral Sir Desmond Dreyer, the Commander-in-Chief, Far East Fleet. Despite the introduction of unified command, this appointment still carried immense prestige and responsibilities. Although the number of ships in the theatre varied, there was usually an impressive array, including an aircraft carrier – probably either the *Eagle* or *Ark Royal* – and its powerful striking force of jet fighters and bombers. But the Commander-in-Chief's place was no longer at sea on the bridge of his flagship – his second-in-command now had that satisfaction – but at GHQ in Singapore, or at Naval Headquarters in the naval base that had been built thirty years before to deter the Japanese. Before the Brunei revolt, the Far East Fleet had mounted some exciting little operations against pirates off the coast of North Borneo but, now that there was a real emergency, the final responsibility for inshore operations had gone to a Gurkha general. What remained for Admiral Dreyer was to meet Walker's requests and to prepare for contingencies such as direct confrontation, or war at sea with Indonesia.

Despite the impressive destructive and theatrical capabilities of his aircraft carrier, Dreyer's fleet was not well-suited to take on the Indonesians. President Sukarno's show-piece, the Russian-built *Irian*, was a Sverdlov-class heavy cruiser, exactly the type which the Fleet Air Arm's Buccaneer bombers had attacked so often in theory and on exercise. But, this apart, there were two threats which the Far East Fleet would find difficult to meet. One was the force of Russian Komar-class patrol boats armed with anti-ship guided-missiles with a range of fifteen miles. Against big ships in confined waters, such as the Indonesian archipelago and the Malacca Straits, these could be extremely dangerous. The Navy hoped that the Indonesians were incapable of maintaining such sophisticated weaponry under such climatic conditions.

The most dangerous threat from the sea was the humble motor-boat; small and powered by a quiet outboard engine. This was how arms could reach the remnants of the Brunei rebels, the CCO in Sarawak and the Indonesians living in North Borneo. Had the Admiralty not disbanded its Coastal Forces there would have been fast motor torpedo-boats and gunboats designed for such tasks. But these had been scrapped – except for a few experimental boats – to balance the cost of a fleet of small coastal and smaller inshore minesweepers to meet the traditional Russian threat of ingenious mine-laying. Now, all that Dreyer could offer to stop the gunrunners were these: slow, under-gunned, their thin wooden hulls standing high out of the water. They could, at least, show the flag and, in the early days, that was important.

Admiral Dreyer had one particular pride. This was the *Albion*. She and her sister-ship *Bulwark* had been light aircraft carriers, transformed into warships of an entirely new concept: the commando ship. In place of her fixed-wing aircraft, each ship carried two squadrons of troop-carrying helicopters, a flotilla of assault landing-craft and one or two commandos of Royal Marines. Assault by helicopter had been developed by the Marines, in the face of a threat of disbandment or, at least, absorption into the Army, and had been successfully demonstrated in the Suez Operation. With a strength of only eight thousand, the Marines felt themselves vulnerable to defence economies and resorted to reinforcing their traditional glamour with abstruse martial expertise. Under-water swimming, rock-climbing, canoeing, commando tactics, beach assaults and then helicopter operations were among these. It was a new concept and one that seemed to fill exactly the needs of a campaign such as was developing in Borneo.

Supposing that Walker was not wholly familar with the commando ship concept, Dreyer set about explaining this. The ship, her helicopters and landing craft and the commandos formed together a single arm and should be used as such. To use them separately would, except in a short period of emergency, be wasteful and, indeed, wrong.

Walker listened attentively. Then he gave Dreyer an unwelcome reply. Under some circumstances, he agreed, the commando ship concept would be ideal and he would certainly use it. But conditions were not now suitable. Operations were in progress a hundred miles from the sea, far too far inland for the Wessex and Whirlwind helicopters to operate from the ship. Moreover, the helicopters were needed there so urgently that not only they and their crews but their ground crews and maintenance equipment would also have to go ashore. And, while the commandos were undoubtedly expert at helicopter operations, there seemed no great mystery about this and other troops could master the skills. So the helicopters would go ashore to be used wherever needed and the commandos would be used where and when they were needed, with or without helicopters. This did not mean that the commando ship would now be wasted. Far from it. She would make an ideal ferry for vehicles, heavy equipment and troops from Singapore.

Admiral Dreyer accepted this news stoically and Walker was to receive excellent support from the Royal Navy. The Fleet Air Arm helicopter squadrons took to jungle operations with such skill, courage and zest that their feats had soon become legendary even in a place so exciting and bizarre as Borneo.

In his dealings with the RAF, Walker faced the dogma of air power and its indivisibility. This doctrine maintained that anything that flew must be considered as part of air power and therefore of the RAF; hence much resentment against the Fleet Air Arm and the dashing little Army Air Corps which was operating light aircraft and helicopters as naturally as other soldiers might drive Land Rovers and armoured cars. Walker saw helicopters, whatever the Service that manned them, as being an extension of the soldiers' role. They were vehicles that happened to fly and therefore should be as closely tied to ground units as, say, artillery.

But the RAF had, according to its doctrine, laid down procedures for Army co-operation. There were structures of command and correct channels for requirements and requests that ensured the indivisibility of control in the air – with the untidy exception of the Navy's and Army's private air forces – and a chain of responsibility going back to the Air Officer Commanding-in-Chief, Far East Air Force.

Walker was as forthright with the RAF as he had been with the Navy. This was to be a helicopter war, he said. In the Borneo jungle, one minute in a helicopter equalled one daylight hour's march; one hour in the air, five days on foot; one battalion with six helicopters equalled one brigade without.

As RAF organisation had been stiff with formalities at the outbreak of the Brunei revolt, so some RAF officers showed an initial inflexibility in Sarawak. There was resistance to unfamiliar ideas and a reluctance to accept risks. When the Fleet Air Arm set up a forward base at Nanga Gaat, on a hillside above the confluence of two rivers some two hundred miles from the sea, it was at a site rejected by the RAF as unsafe for helicopter operations. No such rigidity inhibited the Fleet Air Arm's squadrons and once the RAF began to shake off its conformity in 'going by the book' its air and ground crews began to show similar dash and enthusiasm.

The first AOC-in-C, Far East, Air Vice-Marshal Sir Hector McGregor, a New Zealander, had had differences with the Commander-in-Chief and was replaced by an intelligent and unconventional officer, Air Vice-Marshal Peter Wykeham, a former fighter pilot, an historian and a Socialist whose horizons had been broadened by having J. B. Priestley, the novelist, as a father-in-law. Wykeham did not attempt to interfere with Walker's direction of the campaign and, in any case, he had a major contingency to consider: the possibility of Indonesian air attacks on Singapore and the possibility that he might have to use his V-bombers in retaliatory strikes against Indonesia.

There was always the possibility of a serious clash at sea, particularly when a British warship was in the Sunda Straits between Java and Sumatra on passage between Singapore and Australia. It was British policy to send ships regularly by this route – even when RAF air routes had been diverted far from Indonesia – in order to establish the principle of continuing freedom of the seas. But there were dangerous moments and once when an aircraft carrier was passing the Straits there was real fear of an attack on her by missile-boats and submarines.

Yet there were those in Singapore who would have welcomed the conflict coming to such a head. They argued that this would put a quick end to the war. British contingency plans for 'taking out' Indonesian air power within nine hours and naval power within twenty-four, involved the use of nothing heavier than cannon and rocket fire from strike aircraft. Once in command of the sea and sky, the British could blockade Borneo and, even without further offensive action, the Indonesian war effort would slowly but inevitably run down.

Apart from preparing for this contingency, the RAF had as its main task the logistic support of the ground forces in Borneo. The disputes over command and control of the early days – springing from the old RAF fear of losing its identity – gave way to full co-operation on the insistence of Wykeham. He told his air commanders in Borneo that they must co-operate so efficiently that 'no one else can claim that they could do the job better'.

Wykeham admired Walker as an intelligent soldier whose subtlety on the battlefield was equalled by that in the conference room. But he made a point of keeping on good terms with the Commander-in-Chief, Far East Land Forces, and his staff, often suggesting that they visit Borneo together.

Inevitably, Walker's relations with the Commander-in-Chief, Far East Land Forces, had to be very close. Although, under unified command, and since Walker's appointment to Borneo, this officer had no operational responsibilities in the campaign, he remained the supplier of soldiers and the necessities of war. It was a delicate situation because, while it might be argued that the most simple means of liaison between the general in the field and the general at the base was personal and direct, the chain of command ran from the Director of Operations to the Commander-in-Chief, so by-passing the senior soldier in South-East Asia.

Lieutenant-General Sir Nigel Poett relinquished this appointment early in April. He had been unlucky in being caught by the Brunei revolt not only in the first unsettled days of unified command but while away from Singapore. He had been criticised for failing to send Patterson and 99 Brigade Headquarters to Brunei. But then the internal security of Singapore and the possibility of trouble there coinciding with the outbreak in Borneo must have been on his mind.

His relations with Walker and his circle were not happy. They often seemed to regard him as an amiable peacetime commander without 'fire in his belly' or a tight grasp of operational realities. Walker and he were said to have had acrimonious exchanges and the former could command a scornful tongue. At about this time Walker was asked by a friend to confim the story that when a superior officer had said, in argument, 'You seem to have little confidence in my abilities', Walker had replied, 'You are wrong – I have no confidence in them.' 'It may be true,' Walker said. 'But I have said much rougher things than that.'

In the case of Poett, he seemed convinced that the Commander-in-Chief, Far East Land Forces, not only failed to grasp the vital importance of Gurkhas in the theatre but had also been in connivance with the Army Council's plan for the reduction and possible abolition of the Brigade of Gurkhas. Now, Poett was leaving to retire.

Poett was succeeded by Lieutenant-General Sir Reginald Hewetson. A member of what the Army jokingly called 'the Gunner Mafia' – one of the promising officers of the Royal Artillery whose careers are encouraged and guided by senior Gunners – Hewetson had had a successful career. He had, since the war, commanded both an armoured and an infantry division in Germany before becoming Commandant of the Staff College at Camberley. Since the end of 1961, he had been Commander, British Forces Hong Kong, which amounted to being Commander-in-Chief, although that title belonged, as was the custom in Crown Colonies, to the Governor. In Hong Kong, Hewetson had had important operational responsibilities in what was at the time probably the most sensitive and dangerous point of contact with a potential enemy. He had enjoyed great authority and lived in appropriate style and it must have been galling to be promoted into a supposedly more responsible position only to find it virtually shorn of those responsibilities for which he had been judged fit.

Walker's relations with Hewetson seemed to be no better than they had been with Poett. It seemed to some staff officers that Hewetson was anxious to play a more direct and influential part in Borneo but was firmly put in his place – perhaps with lack of tact – by the Director of Operations.

Hewetson is said to have suggested to Walker that in some purely military matters – the supply of barbed wire for Borneo was thought to have been one specific issue – it would be simpler and more efficient if he put his request direct to Army Headquarters instead of through GHQ. Walker flatly refused to do so, apparently fearing that if he once did this a precedent would be set for what he regarded as unwarranted interference by a single Service.

But Walker felt free to approach Hewetson when he felt the need to do so. Early in the campaign, when the likely tasks ahead in Borneo seemed overwhelming to some of the Servicemen facing the little-known hazards of jungle war, Walker was concerned with morale. He suggested to Hewetson that a special Borneo flash for uniforms and vehicles would be appropriate. Indeed, he had prepared a design. Tom Harrisson had provided a drawing of a Borneo eagle in flight and this was to be superimposed on a shield coloured red and light and dark blue – representing the three Services – and this set on a field of jungle green.

Hewetson replied that he saw no necessity for yet another flash. The 17th Gurkha Division already had the 'black cat' symbol and there were others; quite enough for one theatre. This, it seemed to some, showed that Hewetson was out of touch with the feelings of the fighting men in Borneo and Walker, remarking that he refused 'to do business at this babu level' took his request over Hewetson's head to the Commander-in-Chief, who agreed to it. Relations between the two generals seemed to hover continually on the borders of acrimony.

Walker, as he well knew, was entirely within his rights to refer any matter with any bearing on Borneo operations directly to the Commander-in-Chief, Far East.

But, for many months, Luce had known that in April he was to relinquish command and return home as First Sea Lord. It was important to him that relations between the three Services in the Far East should be as happy as possible because he would surely need all the goodwill he could command for his future efforts on behalf of the Navy in Whitehall. Such fears were to be justified because, a year later, Luce found himself and the Naval Staff facing a ruthless and sometimes unscrupulous attack on the aircraft carrier concept by the RAF. Perhaps because he lacked the quality of intelligent ruthlessness himself, Luce was to lose that battle and, in 1967, he and his Navy Minister, Christopher Mayhew, resigned. The new Commander-in-Chief was to be another sailor, Admiral Sir Varyl Begg. He and Walker had met briefly after the latter had had his interview with General Hull in Whitehall a month before. Walker had immediately liked Begg, sensing in him the qualities of a great commander.

Personally, Begg was a modest man. His brother officers had long expected him to rise to the top of their profession but he seemed to have no such confidence. As a captain, when he commanded the aircraft carrier *Triumph*, he had begun to talk sadly of what he might do when retired and seemed surprised when promoted to Flag rank and, as second-in-command of the Far East Fleet, given one of the Navy's two most important seagoing appointments. He had then become Vice-Chief of the Naval Staff, an appointment that implies further promotion, and now, on 24th April, he took over from Luce in the Far East.

Begg faced a difficult task and he knew it. Like his friend Admiral of the Fleet Earl Mountbatten, the Chief of the Defence Staff, he was a strong advocate of unified command. This had been tried and was working well at Aden in the Middle East Command and had proved its worth in the preventive operation to defend Kuwait against an Iraqi threat. But opposition to the concept from powerful bodies of opinion within the three Services had been as strong in Whitehall as in Singapore. The three Service Commanders-in-Chief in the Far East were his contemporaries and, a month before, had been his equals in rank. In the Far East they had, as he would put it, been 'great taipans' and were restive at being relegated to mere defence advisers and administrative managing-directors. One strongly-expressed view was that instead of having an overall Commander-in-Chief, the three Services commanders should take it in turns to be chairman of a defence committee – but the thought of running a campaign by committee appalled Begg as much as Mountbatten above him and Walker below. There was also strong opposition to Walker's appointment and lobbying by some for his replacement by Hewetson.

Begg arrived determined to make unified command work and to keep the operational chain of command as simple as possible: from himself directly to Walker. In Whitehall he had heard of Walker's reputation both as being by far the most experienced field commander in the Far East and as being a forthright individual. Their liking at the first brief meeting had been mutual and was now quickly to ripen into professional respect.

At the first conference Walker attended at GHQ after Begg's arrival, the new Commander-in-Chief made his attitude clear. Before joining the three Service Commanders-in-Chief, Begg called Walker into his office and asked his opinion of several points they were all to discuss. He listened intently but gave no hint of his own reaction. But when the meeting began it was clear that Begg and Walker not only understood one another and each other's problems but had formed a special relationship that was to remain inviolate.

The future was still a matter for speculation. There were two minor incidents on the frontier with Kalimantan in April; ominous because it was known that Indonesian regular officers had planned and led them. Meanwhile Indonesian opposition to the formation of Malaysia was gathering strength and Sukarno's accusations of an 'imperialist plot' to keep its member-states shackled to Britain made political refutation as important as military preparedness. This was primarily the responsibility of Lord Head, the Governor-General in Kuala Lumpur, who would become British High Commissioner at the formation of the Federation. Here again the choice had been a happy one, giving Begg the reassurance that the political aspect of the campaign was in capable hands.

The capture of Yassin Affendi on 18th May marked the final ending of the Brunei revolt and now all three Borneo territories presented a variety of dangers to Begg as he discussed contingency plans with Walker. The theatre of war was vast and wild. Together, Sarawak, Brunei and North Borneo covered an area as large as England and Scotland. The frontier with Indonesian Borneo, or Kalimantan, ran for nearly a thousand miles mostly along the top of the major watershed range rising to a height of eight thousand feet. Most of the hinterland was primary jungle with a treetop canopy as much as two hundred feet from the ground, dense secondary jungle, where the virgin growth had been felled, and wide belts of virtually impenetrable mangrove swamp along the coast. There were a few roads in the immediate neighbourhood of the towns, rivers providing the main means of transport, the greatest of these being the Rejang in central Sarawak which was navigable by coastal steamers as far as the town of Kapit more than one hundred and sixty miles from the sea.

To defend all this area only five battalions were available, one of them, the Royal Leicestershire Regiment, having come down from Hong Kong with two others during the revolt as an immediate reserve and stayed behind when the others returned. This force came under one brigade headquarters in Brunei and another at Kuching with the Director of Operations presiding from Labuan Island.

With so small a force available, Walker had to rely first on intelligence, and, secondly, on helicopters for mobility. The Special Air Service squadron, which had arrived early in the year, was relieved early in the summer by another, which came almost directly from exercising in Arctic Norway and the headquarters of the regiment arrived from England to take direct control of operations on the frontier.

The SAS four-man patrols along the border had proved their worth but there was a need for a finer mesh in the surveillance net. This could only be achieved by increasing the patrolling and intelligence-gathering skills of the infantry battalions. A way of doing this was noted by Walker when visiting a patrol base manned by Royal Marines in the First Division of Sarawak. He had arrived there from Miri where John Fisher, the Resident, had been suggesting that a further use might be found for Harrisson's Force in watching the border and longhouses in vulnerable positions. Now he discovered that the commandos had had the same idea. He was impressed by the detail of their local topographical knowledge and, asking how they had acquired it, was told that it was due to 'Tom Thumb', a Bornean guide.

The Marines had given Tom Thumb jungle kit, a shot-gun and rations and, in return, he had guided their patrols and had given warning of an imminent raid from Kalimantan. There were plenty of potential guides in the longhouses, the Marines told Walker, but it was advisable to avoid what they called the 'Flash Alfs' who had been educated at Kuching.

On returning to his headquarters Walker suggested raising a force of Bornean irregulars, partly from Harrisson's Force, pointing out that as the Sarawak Rangers had been used successfully as trackers in Malaya, they should be even more effective in their own country. The plan was first rejected by the senior police officers in Sarawak as being likely to produce only 'third-class soldiers'. But Admiral Begg supported Walker and, in May, the Sarawak Government agreed to the raising of an initial three hundred 'auxiliary police' to be known as Border Scouts.

The recruits were trained by the SAS and the Gurkha Parachute Company in basic tactics and weapon-handling and issued with jungle-green uniform and light-scale equipment. Soon after, the raising of more Border Scouts was authorised to a total of about one thousand.

The problems of anticipating and intercepting border raids and containing internal unrest were thrown into sharp relief in May when Walker's central intelligence team – including an authority on Indonesia – were killed in the crash of a Belvedere helicopter. At moments the outlook seemed almost desperate, so heavily did the odds seem weighed in favour of the enemy and his aggressive plans.

The greatest single threat was internal and buried like an explosive charge of unknown strength deep in the polyglot population of the Borneo territories. In a total population of just under one and a half million, only about half were indigenous. In Sarawak, thirty-one per cent of the people were Chinese and nineteen per cent Malay; in Brunei, fifty-four per cent were Malay and twenty-eight per cent Chinese; in North Borneo twenty-one per cent were Chinese and seven per cent Malay. Most of this massive Chinese population had settled in or near the towns or at trading posts or on farms beside the major rivers. The indigenous tribes – the Land Dyaks, Ibans, Kenyahs, Kayans, Muruts, Kelabits and the nomadic Punans – were, for the most part, unsophisticated, agricultural people living in tribal longhouses or kampongs. Here, as in Malaya, the industrious, ambitious Chinese minority were dominant.

The Clandestine Communist Organisation in Sarawak was estimated to consist of a hard core of about one thousand Chinese activ-
ists in some two hundred and fifty cadres, a further fighting strength
of three thousand five hundred and as many as twenty-five thousand
sympathisers and potential fighters. Links with Indonesia had been
strengthened as result of the otherwise successful rounding up of
weapons by the Army and Police in April because as many as a thousand young Chinese were known to have then crossed into Kalimantan for weapon-training with the Indonesian Army.

In Brunei there were still many sympathisers of the rebels in the
capital and the villages and, although the TNKU had been destroyed,
there was the fear that the Sultan's own bodyguard, the Brunei Regiment, might, in the tradition of palace revolutions, itself become the
vanguard of a new insurgency.

Finally there were some thirty thousand Indonesians living in
North Borneo, two-thirds of them in the Tawau district. While there
seemed no immediate threat here, Indonesian intelligence agents
had been active there and were known to have drawn up plans for
sabotage and ambushes.

While this potentially hostile force of about sixty thousand Chinese and
Indonesians were the powder-keg, the fuse lay across the border. These
were the Indonesian Border Terrorists – the IBT – who were believed to
number one thousand five hundred and be supported by an unknown
number of Indonesian regular soldiers and local defence irregulars.

These were deployed along the whole length of the frontier under
the overall command of an Indonesian officer, Lieutenant-General
Zulkipli. Of their eight operational units, only one faced North Borneo and the Fifth Division of Sarawak. The strongest concentrations
were thought to be opposite the First and Second Divisions and the
tempting targets of Kuching and Sibu. Led by Indonesian officers, the
units were trained in commando tactics and boasted such aggressive names as the Thunderbolts, the Night Ghosts and the World
Sweepers.

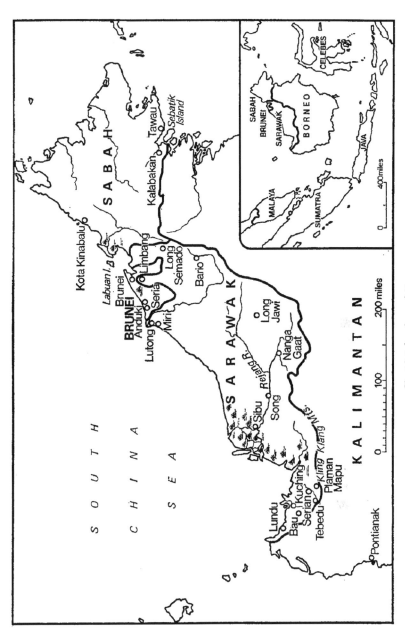

East Malaysia and Borneo

To meet these combined threats Walker had his five battalions as a mobile defence force. But this mobility was severely limited by the small number of aircraft available to lift them. Throughout the summer of 1963 this generally amounted to about a dozen helicopters – each able to lift between ten and sixteen fully-equipped infantrymen together with one large Beverley transport, two or three medium Valettas and about four Twin Pioneers with a total capacity of about ninety, thirty and a dozen armed men, respectively.

Despite apprehensions, the four months following the Tebedu raid were relatively quiet. At Walker's headquarters this was interpreted as a period of preparation by the enemy which had to be countered by intensive planning, training and defensive operations. But there were others, notably in Singapore, who foresaw no more than political opposition and perhaps token demonstrations of force as preparations for the foundation of Malaysia went ahead. Walker was already being accused of 'crying "Wolf!"'

Political confrontation there already was and this was intensified by support given to President Sukarno by President Macapagal of the Philippines, which had an historic claim to sovereignty over North Borneo and so readily joined the opposition to the new Federation. While there seemed no risk of a Philippino attempt to annex North Borneo or to take military action in support of subversion there, Macapagal appeared ready to join Sukarno in his grandiose plans for a Maphilindo empire. This concept had a romantic appeal in South-East Asia for it involved the cutting of the last imperialist links with Europe and the United States and could be presented as a renaissance of the golden age which plays so potent a part in Asian historical mythology.

While the Tunku Abdul Rahman, as the President-elect of Malaysia, had no wish to ally Malaya – with the highest standard of living in South-East Asia – with the potentially rich but bankrupt and erratically-led Indonesia or with the politically-disturbed Philippines, he felt he must attempt to find grounds for a rapprochement. He agreed first to a meeting of the three countries' foreign ministers at Manila in June and, two months later, a 'summit conference' there of the three heads of state. At this the Tunku agreed that an 'impartial' team of United Nations observers should visit Sarawak and North Borneo immediately to assess public opinion and the popularity of the concept of Malaysia.

Up to this time Borneo had experienced little more than the spasmodic troubles of the frontier. Following two raids on border posts in Sarawak, which had inflicted no casualties, a five-mile belt along the frontier in the First and Second Divisions had been declared a curfew area giving the defenders freedom to fire at will during specified hours of darkness. This reduced terrorist activity to small raids on border longhouses and trading posts well away from patrol bases. However, towards the end of June, a party of about fifteen uniformed guerrillas crossed the border in the Second Division in an attempt to attack longhouses and perhaps a post manned by the newly-formed Border Scouts. Police agents reported their presence and the 2/10th Gurkhas with only one Whirlwind helicopter to give mobility set a series of traps, in one of which Border Scouts were used as bait. Operations continued for nearly a fortnight, the Gurkhas making up for their lack of helicopters by imaginative tactics. Two of their ambushes succeeded, two enemy were killed and the guerrillas faded away into the jungle on the frontier.

Two principal lessons were learned from these skirmishes. One was that so long as raiding parties could range freely up and down their side of the frontier probing for undefended targets – this party had travelled forty miles along the border from their base – they enjoyed an immense advantage. The other was that speed and flexibility were the battle-winners for the defenders: on two occasions it had taken the Gurkhas six hours or more to reach the scene of a reported incident. Walker ordered that a Whirlwind with a section of Gurkhas should be kept at instant readiness for take-off and that reaction time must be reduced from hours to minutes.

It seemed inevitable to Walker's staff that the Indonesians would take some action to create unrest before the United Nations team arrived towards the end of August and the pattern of their activity suggested that it might be a deep-penetration raid with the objective of triggering riots or even a localised uprising by the CCO. As usual, Kuching and Sibu were regarded as the two danger-points but, across the frontier, General Zulkipli, realising that the defenders there would be particularly alert, chose to put in an attack where it would be least expected.

On the evening of 14th August, a Pengulu – an Iban headman – reported to the District Officer at Song, a small town on the great Rejang river, some thirty miles from the Indonesian border and about seventy miles from the sea, that raiders had crossed the frontier to the south and had seized four of his people.

The Third Division, which offered no obvious targets to raiders, was the responsibility of the 2/6th Gurkhas, who were stretched thinly as far as Tawau on the coast. A patrol was sent to investigate and, next day, they sighted longboats filled with refugees hurrying down a river flowing from the direction of the border. Among these were the four men who had been seized by the raiders when they had first appeared and they gave an alarming account of the enemy. There were, they said, about fifty of them, well armed and wearing uniform and jungle equipment. They were led by officers and non-commissioned officers who had told them not only that their aim was to kill white men and keep the Malays out of Borneo but that they were the spearhead of heavier attacks. Three hundred more raiders were due to invade within a week and another six hundred within a fortnight.

Next day a Gurkha patrol made contact with the enemy and a fierce fire-fight broke out. The Indonesians seemed to include some well-trained soldiers and to be professionally led but their expenditure of ammunition was high and their fire-control deteriorated as the battle continued. During redeployment in an attempt to outflank the enemy, the British patrol-commander became separated from his men and was seriously wounded. Unable to move, he was found by the Indonesians and, after putting up a brave fight with his sub-machine-gun, was killed.

Meanwhile another Gurkha patrol had found more enemy and there began what was to become a familiar pattern of patrols and ambushes being put into position by helicopter like chess-men to counter what might be the enemy's next move. It was known that the Indonesians had suffered some casualties but where they were and whether they had been reinforced could only be assessed from scraps of intelligence and disciplined speculation. Yet contact was made every few days and more casualties inflicted. On 31st August, six were killed and, three days later, two, including the leader of the invaders. A month after they had first been sighted, the Indonesians had lost fifteen killed and three captured.

During the battle the 1/2nd Gurkhas flew from Singapore to relieve the two companies of the 2/6th, which had been involved earlier but now closed up with the rest of the battalion in East Sarawak. This had been the first deep-penetration raid but it was likely that others would follow and now it seemed that the enemy was as likely to choose 'soft targets' in less important and lightly-defended areas as in the obvious danger zones.

Interrogation of prisoners now threw fresh light on the raid. The invaders, led by Indonesian regular officers and NCOs, were attempting to capture the riverside town of Song. But they had crossed the border without maps or compasses, relying on local guides, and, without rations, hoping to live off the land. Their aim was so over-optimistic that it could be assumed either that their intelligence assessment of unrest in Sarawak was wildly inaccurate, or that the party had been sacrificed as a propaganda exercise to impress the United Nations' observers who were arriving at the end of August. The latter seemed the most probable explanation and reports of the actions and the killing of the British officer did, of course, make far more impact in the news than had any move since the Tebedu raid.

The United Nations team, accompanied by Indonesian and Philippino observers, had arrived and set out to tour Sarawak and North Borneo. They confirmed that the recent elections, which were, in effect, a referendum on the Malaysia issue, had been fairly conducted although there was a large Chinese minority solidly against the prospect of the Federation. In North Borneo there had been little opposition to the plan.

When they had reported to the Secretary-General of the United Nations, U Thant confirmed that 'the majority of the peoples of the two territories ... wish to engage with the peoples of the Federation of Malaya and Singapore in an enlarged Federation of Malaysia ...'. This, combined with the Indonesian observers' agreement that the United Nations team had acted with impartiality, might have taken the wind out of the sails of any opposition other than Sukarno's. As it was, President Macapagal was provided with a welcome excuse to become even less enthusiastic over a Maphilindo empire, contenting himself with platitudes about Asian affairs being run by Asians.

But Sukarno was not to be put off, for his Maphilindo dream had expanded to include Burma, Siam, Laos and Cambodia. This, wrote a newspaper columnist in Manila, would have to becalled Maphilindosiburmlicam, which sounded more Welsh than Asian. Clearly he would do all he could to prevent the birth of Malaysia from being easy.

On 16th September, the Federation of Malaysia was established. In Borneo, the British Governors of Sarawak and North Borneo became High Commissioners and the latter state changed its name to Sabah. Brunei, however, remained a British protectorate despite British urging that the Sultan should change his mind and join Malaysia at the last moment. But while the Foreign Office and Commonwealth Relations Office deplored his intransigence, there were British commanders in South-East Asia who welcomed his stand on the practical grounds that it gave Britain a military foothold in Borneo that was not subject to Malaysian political influence, so giving them an added measure of independence.

The creation of Malaysia brought about immediate change in the command structure to which Walker was answerable. Overnight what had been a colonial war was transformed into the support of a Commonwealth ally. Now the supreme military authority was no longer GHQ at Singapore but the Malaysian National Defence Council in Kuala Lumpur.

This was the policy-forming body, sitting under the chairmanship of the Prime Minister, linked with London through the British High Commissioner, Lord Head, and including, when necessary, the British Commander-in-Chief, Far East. Implementing the Council's broad strategic decisions was the National Operations Committee with the Chief of the Malaysian Armed Forces Staff, General Tunku Osman, and the Inspector-General of Police, Claude Fenner, as Joint Chairmen and with Admiral Begg as a member through whom instructions were given to Walker.

There was also limited political control within Borneo. In both Sarawak and Sabah there were State Security Executive Committees both answerable to the National Operations Council but with Walker as a member of each. He was also a member of the somewhat similar State Advisory Council in Brunei which was answerable to the Sultan.

The machinery of command was cumbersome but, largely due to Begg's determination and tact, it worked effectively. One obstacle was to remain that of the command and control of police in Borneo. Although Walker could exercise operational control over those allocated to him, the police in Sarawak and Sabah had their own command structure with its highest authority in its Inspector-General. Sometimes, it was thought, Fenner acted as though he felt that he, with his experience of the Malayan Emergency, should become Director of Operations in Borneo. Certainly, with the support of the Tunku, he urged Walker to move his headquarters to Kuala Lumpur or, failing that, Kuching. Walker refused, saying that he was fighting a war on a thousand-mile front, not conducting a police internal security operation, and only from his headquarters on Labuan could he stand back and survey the whole theatre and reach any point with equal ease.

Despite disagreements and the unsatisfactory system of the police being under separate command from the soldiers, Walker and Fenner remained personal friends and, when Fenner's attitudes seemed to be alienating Begg, it was Walker who brought them together and helped create a friendly relationship. For his part, Walker would have liked to see the weight of the command structure moved further forward for now he had to make regular – usually fortnightly – journeys to Kuala Lumpur as well as to Singapore.

Walker's thoughts were concentrated on contingencies as Malaysia's 'Freedom Day' approached. The event itself was to be marked by a visit from Mr Duncan Sandys, the Secretary of State for Commonwealth Relations, to the new East Malaysia. On this occasion Walker was in two minds about the advisability of accompanying his guest on his tour because this would keep him away from his headquarters at a time when any emergency, singly or in combination, could arise. On the other hand, Sandys himself might provide the flash-point of disturbance and it might, after all, be prudent to accompany him. Also, Walker had heard the Minister's reputation as something of a bully and, on his arrival at Brunei, he was struck by the force of his character. So another argument in favour of travelling with him was to act as a buffer between him and the brigade and battalion commanders he wanted to visit.

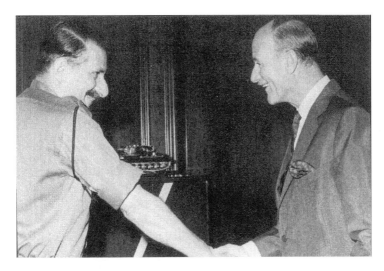

Walker arrives. Reunion in New Delhi with General Sam Manekshaw, Commander-in-Chief of the Indian Army in 1970

Hull departs. Field-Marshal Sir Richard Hull retiring as Chief of the Defence Staff in 1967 makes his farewell to Denis Healey, the Defence Minister *(right)* and Admiral Sir Varyl Begg, the First Sea Lord *(centre)* outside the Ministry of Defence in Whitehall

NATO on parade. The Commander-in-Chief, Northern Europe, and his wife (*left*) watch a military display at Ålesund, Norway, accompanied by senior Norwegian officers including Lieutenant-General H. Løken, the Commander-in-Chief of the Army *(second from right)*

NATO *en fête*. Walker talks with Lieutenant-General O. Blixenkrone-Møller, Commander-in-Chief of the Danish Army, at a ball in Frederiksborg Castle in 1971

Walker accompanied Sandys – they were later joined by Begg – and was impressed by his realistic views. He also found that he need not have worried about his impact on the battalion commanders after one of them politely rebuked Sandys for interrupting his briefing, asking him to wait for the question period at the end.

But Walker was disturbed at words shouted by Sandys against the roar of aircraft engines during one of their flights. 'It is not the policy of Her Majesty's Government to become caught up in a war,' he said. 'Try to stop it from escalating. Do everything you can to stop it.' Later Walker confided that the way this had been put smacked of defensive thinking. The war had already started and the emphasis should be on ending it by winning.

Indonesian reaction to the foundation of Malaysia was immediate and furious. The Malayan ambassador in Jakarta was expelled and the British Embassy sacked and burned. Sukarno denounced the new Federation in hysterical language, promising it 'a terrible confrontation'.

War was not declared but now the Indonesians undertook the most warlike action that had yet been seen in Borneo. Again, their chosen target was not one of the tempting towns. This time it was perhaps the most remote and deeply-buried settlement in what was now called East Malaysia, more than a hundred miles from the sea. A raiding party of about two hundred Indonesians supported by about three hundred unarmed porters crossed the mountains on the border into the wild and roadless Third Division, penetrated more than fifty miles down the Balui river in longboats and fell upon the unsuspecting outpost at Long Jawi.

This was a village on a river trade route between the heartland of Borneo and the great Rejang river, the port of Sibu and the sea. There were about five hundred inhabitants, mostly belonging to a tribe which spread both sides of the border and, until this September, they had been remote from the war. However, as Long Jawi lay at the junction of river and track communications, the 1/2nd Gurkhas had a position there. Based on the schoolhouse, which they lightly fortified, were six Gurkhas, three policemen and twenty-one Border Scouts and the force included a Gurkha and a Police wireless-operator. On 27th September, one of their officers flew in for an inspection and found all quiet. There was no hint that for the past two days Indonesian raiders had been infiltrating the village.

At five o'clock next morning the attack was launched. The first burst of machine-gun fire raked the schoolhouse and mortar bombs scored a direct hit which knocked out both wireless sets and killed their operators. Three others – a Gurkha, a policeman and a Border Scout – died in the firing and the rest of the defenders retreated into the jungle where some of the Border Scouts were taken prisoner. Long Jawi had fallen and the Indonesians had scored a real military success.

Because wireless communication had been cut, news of the attack did not reach the battalion headquarters of the 1/2nd Gurkhas for two days. I reached Lieutenant-Colonel John Clements – a veteran of Imphal – by three Border Scouts who had managed to escape. This was the repetition of the attempt on Song which Walker had anticipated and he assigned all the Wessex helicopters of 845 Squadron, Fleet Air Arm, to Clements, who at once flew Gurkhas forward to take up positions as 'stops' on the enemy's probable line of retreat.

This was rightly assumed to be the Balui river for, that same day, a longboat travelling upstream was ambushed and three armed Indonesians on board killed. Next day, helicopters flew a platoon to another ambush position on the river bank, the Gurkhas sliding down ropes from the hovering aircraft. That evening the Gurkha lieutenant commanding the ambush heard approaching outboard motors and soon, through the jungle leaves, he saw two longboats coming up-river, each crowded with armed Indonesians. The Gurkhas waited until the boats reached the arranged killing-ground, then opened fire. One boat sank, the other ran aground and twenty-six Indonesians died in a few seconds.

The formal chess game of planting stops and sending out patrols continued. Twelve days later, the Gurkhas discovered an abandoned enemy camp, five dead Indonesians and the remains of seven captured Border Scouts, who had been tortured and killed. Ambushes continued, with occasional success, until the end of October when it was believed all the Indonesians not accounted for had escaped across the frontier, or had wandered into the Sarawak jungle to starve.

The tables had been turned on the invaders. After their initial surprise attack and the infliction of five killed and seven prisoners murdered, they themselves had suffered thirty-three killed and perhaps others lost in the jungle. Walker had learned several lessons, the most vital of which was from the surprise the enemy had achieved, although the Long Jawi villagers had known of their presence for two days. This underlined the importance of establishing close relations with the tribes through personal contacts and the 'hearts and minds' campaign. Also, the Border Scouts had failed to give warning, so failing to fulfil their primary task as 'eyes and ears'.

The failure of the Border Scouts at Long Jawi led eventually to a change in their role. Walker decided that they should concentrate almost entirely on intelligence work at the expense of their para-military duties. Thus most of them returned to their villages in civilian clothes and carried on with their original occupations, passing on any reports of Indonesian activity and any significant rumours to the security forces. Many of them had friends and relations who regularly crossed into Kalimantan and their reports, verified by SAS and other surveillance patrols, were to provide Walker with the sensitive mesh of watchers and listeners that he needed.

There was a change in Walker's command structure at this time. Brigadier Glennie returned to Borneo as Deputy Director of Operations and to command in Brunei and Sabah, while Patterson's 99 Brigade Headquarters returned to Singapore. Sarawak remained the responsibility of 3 Commando Brigade Headquarters at Kuching. The command structure above Walker and the Commander-in-Chief had also changed and this was to be reflected on the battlefield. Immediately after the founding of Malaysia, Malaysian soldiers joined the British and the Gurkhas in the Borneo jungle.

Most new armies face problems and learn lessons the hard way on first going into battle and the Malaysians were no exceptions. Those of them who had served in the Royal Malay Regiment under British officers in the Malayan Emergency had experienced active service and some of these had made excellent NCOs. Now most of the British officers had left and their places taken by Malays, most of them young, inexperienced and enthusiastic. There was potential military talent in the Malaysian armed forces but with their sudden expansion this was inevitably spread thin.

Among signs of health in the Royal Malay Regiment battalions, which were quickly sent to Borneo, was pride and an eagerness to be given the positions of danger and honour. Thus, the 3rd Battalion went to Tawau, where the threat from Indonesian regulars and the large Indonesian immigrant population was real, and the 5th went to the First Division, at the opposite end of the front, to join the Royal Marines defending Kuching.

The Malays showed some of the failings of young, inexperienced soldiers anywhere. The business of going off to war seemed romantic, particularly now that they were defending what some of them seemed to regarded as their very own empire. Their pride was reflected in their dress. The baggy, jungle-green denims and round, wavy-brimmed hats with which they were issued – clothing designed specifically for ease of movement and camouflage in the jungle – were not to their liking and they tailored their shirts to fit their chests, tapered their trousers and cut back the brims of their hats, binding the crowns with their regimental colours. They were fastidiously clean, using talcum powder and hair oil and totally disregarding Walker's instructions that soldiers in the jungle must live like animals. The Gurkhas said that they could smell the Malays' toiletries a mile away.

In the excitement of going on active service the 3rd Battalion overlooked some basic, practical precautions. They had, in fact, been told what to expect by Brigadier Glennie himself. An enthusiastic yachtsman, Glennie took a particular interest in the maze of creeks and rivers that ran through the swamps along the Tawau coastline and had himself set about organising a flotilla of small patrol craft that came to be known rather grandly as the Tawau Squadron. He studied the particular problems of this end of the Borneo flank and decided that an attack was to be expected.

Tawau was almost as tempting a target as Kuching. It was the prosperous centre for many rural industries – timber, rubber, tea, cocoa, hemp and palm oil – and three-fifths of the population were Indonesian immigrant workers and their families. The port itself stood on the northern shore of Cowie Harbour, and across the bay lay Sebatik Island, some thirty miles long, of which half belonged to Malaysia, half to Indonesia. On Sebatik and along the rest of the frontier the Indonesians had deployed five companies of regular marines – the *Korps Komando Operasi* or KKO – and a training camp for volunteer terrorists, some of them from the logging camps of Sabah.

The 3rd Battalion, Royal Malay Regiment, together with a company of the King's Own Yorkshire Light Infantry and some local gendarmerie, were responsible for the defence of Tawau and its environs and Glennie told them to expect an attack. Despite the presence of the Indonesian marines he did not expect this to come by sea, if only because the Royal Navy always kept a frigate or destroyer off Tawau and an efficient radar watch was maintained. But he did think that the attack would come through the swamps and he advised the newly-arrived Malays to prepare for it with patrolling and a defence system in depth.

They had not acted on Glennie's advice when a strong force of raiders – thirty-five Indonesian regulars leading one hundred and twenty-eight volunteers – crossed the border. Their intention was first to capture the defended village of Kalabakan to the west of Tawau and at the head of Cowie Harbour, then, as the Indonesian expatriates rose to support them, advance on Tawau itself. After crossing into

Sabah, the raiders lurked undetected in the swamps and forests for eight days, then on 29th December were ready to launch their attack.

There were two defensive positions at Kalabakan. On the bank of the river running into Cowie Harbour stood the police station, fortified with sandbags and barbed wire and manned by fifteen policemen. Four hundred yards away, a platoon and two sections of the Royal Malay Regiment, under their company commander, occupied two huts. Although trenches had been dug nearby, the huts themselves were not fortified in any way. After dark that evening, the unsuspecting Malay soldiers were engaged in domestic activities, washing and cleaning clothes and equipment. Their sentries were not alert.

Meanwhile, two hundred miles away at Brunei, Walker was listening to news of another war. Brigadier Frank Hassett, who had recently commanded the 28 Commonwealth Brigade in what was now West Malaysia, had just arrived from Vietnam. He was an Australian and because the South-East Asia Treaty Organisation, of which both Australia and the United States were members, was not involved with the American commitments in Vietnam, he had been officially forbidden to go. But on his way back to Australia after attending the Imperial Defence College in London he managed to visit Saigon privately and unofficially when on leave. Now he was giving Walker a first-hand account of a war that had escalated to a point that might well be reached in Borneo during the following year.

Hassett was staying with the Walkers at Muara Lodge, the Sultan's beach house on the coast some thirty miles from Brunei Town. The Sultan had offered it to Walker as his official residence when, a year after his arrival in Brunei, Beryl had been allowed to join him. There, in his air-conditioned study, he could work undisturbed far into the night and there he could meet the Sultan for private talks without arousing suspicions that they were becoming too friendly for British or Malaysian susceptibilities. The house appeared relaxed and, indeed, vulnerable because Walker kept his Gurkha guard hidden away and, emulating the Japanese, kept a Gurkha with binoculars and a machine-gun hidden in a tree.

Borneo seemed quiet on the day Hassett arrived, although when Beryl took him for a swim they saw a suspicious figure on the beach. The guard was called but nothing further seen and the incident became another subject for conversation at dinner. Hassett was tired after his journey and went to bed early, Beryl giving him sleeping pills to ensure a restful night. He was the only person at Muara Lodge who enjoyed that.

At two o'clock in the morning Walker's telephone rang and the duty officer at his headquarters told him that Kalabakan had been attacked and had fallen.

The attack had been launched just before eleven o'clock. An assault group had crept up to the Malay huts, flung grenades through the windows and opened automatic fire. Before the Malays could reach their weapons, eight had been killed, including the company commander, and nineteen wounded. Ten minutes later, another party had assaulted the police station but the small garrison had had time to rush inside their perimeter wire and successfully fought off the attack. Now the Indonesians raced through the village and the logging camp. In the timber company manager's house they helped themselves to his whisky, while his wife and their cook hid under a bed. The manager himself, out at a logging camp, made for the Malays' strongpoint. Reaching the survivors, he told them that this was the moment to counter-attack and he would join them himself. But, shocked into inaction, they stayed where they were.

The Indonesians moved out of the village but, content with their success, did not move directly upon Tawau, where they might have been able to repeat it. Instead, they travelled north to lie low in the wilds where they believed they would be safe.

Next morning Walker flew to Tawau, taking the surprised Hassett with him not as a personal guest but as a useful observer. The night before, Hassett had said that some official Australian opinion held that the Indonesian confrontation of Malaysia was bluff and that reports of imminent war in Borneo were wildly exaggerated. Walker hoped that Hassett would see that what had happened at Kalabakan was no less than war and would report to Canberra accordingly.

On arrival, Walker sent a signal to Begg asking for urgent reinforcements and, at once, the 1/10th Gurkhas, who had already served in Sarawak, began returning from Malacca. But Walker also heard that the news had been reported to the Tunku Abdul Rahman verbally and, to spare his feelings, the signal giving details of the action had been kept from him. On hearing of the disaster he had at once announced his intention of flying to Sabah to visit the survivors of the first battle fought directly between Malaysia and Indonesia.

Walker flew to Jesselton – the capital of Sabah, now to be renamed Kota Kinabalu – to meet him and was relieved when Fenner arrived by an earlier aircraft. Walker told him that the inexperienced Malay soldiers had been 'caught with their trousers down'. This often happened under such circumstances and was a lesson to be learned. He intended to tell the Tunku exactly what had happened. Fenner strongly advised him not to do so. Whether or not British troops had suffered similar initial disasters in war, the Tunku would never accept or believe that his soldiers had been caught by surprise and defeated. Walker, realising that Fenner knew the Tunku better than he did, agreed.

When the Tunku arrived, Walker told him the story of the action in diplomatic language. The Indonesians had approached with great skill, using the cover of Kalabakan village, so enabling them to rush the Malay position in overwhelming numbers. This was both true and tactful. The Tunku was intensely moved and, on arrival at Tawau, tearfully greeted his wounded soldiers, gave each a handsome cash bounty and decreed that a monument should be erected at Kalabakan to the dead.

Meanwhile the Gurkha reinforcements were flying in and assembling at Tawau under their commanding officer, Lieutenant-Colonel E.

J. S. Burnett. In close co-operation with Glennie's armada of patrol craft, which now formally became the Tawau Assault Group and included a raft mounting a three-inch mortar and named the *Monitor*, the Gurkhas began a meticulous programme of patrolling and ambushing, using all the skills of their trade and living, as Walker had hoped all his soldiers would, like beasts of prey.

About a month later, it was over and all but six of the enemy had been killed, captured or had surrendered. Like the raid on Long Jawi – and like so many battles and wars in which the British had been engaged – this had opened with an easy victory for their enemy but the tables had been turned by a combination of resolution and professional expertise. Colonel Burnett was awarded one of the two Distinguished Service Orders won in Borneo.

Again there was overwhelming evidence that Indonesian regulars had taken part in attacks over the border. Two-thirds of the KKO marines involved were killed or captured and the prisoners admitted that they had been sent into East Malaysia expecting that the population would rise and greet them as liberators.

The initial success of the Kalabakan raid rebounded on the Indonesians because it at once changed what had been proclaimed as an 'anti-imperialist' crusade into a war between Asians. Certainly, after this time, most Malays in Malaysia felt themselves more fully involved than hitherto. The Indonesians made attempts to shrug off direct responsibility and claimed that the KKO marines who had taken part were enthusiastic idealists who had resigned in order to join the volunteers. In Kalimantan, the Indonesian Foreign Minister, Dr Subandrio, produced Azahari, the absentee leader of the Brunei revolt, to repeat that Indonesian regular forces were playing no active part in the Borneo operations.

This contortion proved to be the opening move in a new political offensive which Sukarno was preparing, but, meanwhile, border incursions continued. A notable aspect of the actions fought during January, 1964, was the extent to which British infantry became involved.

Hitherto, most of the stiff fighting had fallen to Gurkhas but now it was the turn of the British to prove themselves.

The 1st Battalion, the Royal Leicestershire Regiment, commanded by Lieutenant-Colonel Peter Badger, was the first to fight. Patrolling in the Fifth Division at the beginning of the month, they had found an Indonesian machine-gun post near the border, destroyed it and killed the four-man crew. Three weeks later, a Border Scout reported a recently-abandoned enemy camp which could have been occupied by at least five hundred men. Undaunted by numbers, a fighting patrol of eighteen Leicesters and two Border Scouts, led by a second-lieutenant, followed the enemy's trail. On the 24th, they found another empty camp, which appeared to have been occupied by about eighty men, and the tracks leading from it towards the Kalimantan border were only one day old. The Leicesters shed their packs, ready for instant action, and moved forward, their Border Scouts deciding to stay where they were.

Farther along the track, the roofs of two huts were sighted and they were clearly occupied. While a cut-off party was working its way round to the far side of the camp, the Leicesters were sighted and immediately their officer gave the order to charge. Firing from the hip, eleven men rushed the enemy, racing through the camp, then turning to occupy it and await a counter-attack. But the Indonesians had fled into the jungle, leaving seven dead. The camp had been occupied by about sixty raiders and they had left behind more than half a ton of weapons and ammunition, equipment, uniforms (including a peaked cap with a silver star badge) and documents. It was a small, complete victory.

Another significant action took place that month at the far end of the front when about one hundred well-armed raiders crossed into the First Division with the apparent aim of attacking Kuching airfield. They were intercepted and put to flight by a small patrol of Royal Marines and local gendarmerie after a sharp fight in which they lost two killed and several wounded and the brave commando corporal, who commanded the defenders, was killed.

Ominously, among the weapons captured were two rocket-launchers, one of them Czech-made, and a light-weight, high-velocity Armalite rifle. This latter was an American weapon which British trials in 1961 had been shown to be ideal for jungle warfare. Yet Walker's attempts to replace his soldiers' heavy NATO rifle with this had only resulted in a small number being sent to Borneo for further trials.

Jungle operations were in progress along the length of the frontier when Sukarno launched his 'peace offensive'. Indeed, the Leicesters' patrol, which was out of wireless communication with its base as its quarry was also, stormed the raiders' camp some hours after reports had reached Brunei and Singapore that Sukarno was ready to agree to a cease-fire – so, by implication, contradicting his claim that Indonesian forces were not involved in the fighting.

The Indonesian move was in response to an appeal by the Secretary-General of the United Nations that the Governments of Indonesia and Malaysia should agree to meet for peace talks. The cease-fire was announced in Jakarta on 23rd January and was to take effect two days later. Walker suspected this as being at best insincere, at worst a trick. On receipt of Sukarno's offer, the National Operations Council had ordered their Director of Operations to ease the pressure and, while trying to prevent further infiltration from Kalimantan, to allow infiltrators already over the border to return peacefully and, in operations already in progress, to try to capture rather than kill the enemy.

Yet violations of the frontier continued while ministerial delegations from Indonesia, the Philippines and Malaysia met in Bangkok in February. The talks, which had been welcomed by Mr Robert Kennedy, the United States Attorney-General, who was touring South-East Asia, never seemed likely to succeed. They began with no agenda and much bombastic posturing and served only to harden attitudes on both sides.

In Borneo, meanwhile, nothing seemed to have changed and crossings of the frontier continued on land and in the air – Indonesian fighters and bombers had been deliberately 'buzzing' towns in

Sarawak for about three months – and, on 24th February, the Malaysian National Operations Council set up an Air Defence Identification Zone. A fortnight later they ordered Walker to resume operations using 'all possible means' with a view to 'eliminating all Indonesian border terrorists in Malaysian territory'.

Walker had not expected the cease-fire to lead to peace and had written: 'No one knows where this exercise in brinkmanship will end. We are sure only of one thing: we have set our faces to the enemy, and, until more reasonable counsels prevail, we shall not look back.'

When full-scale operations were resumed, a change was seen to have taken place in the enemy's strategy and tactics. Now their incursions were far more professional, they fought with skill and tenacity and the fighting became less like fighting communist terrorists in Malaya and more like fighting Japanese soldiers in Burma. What had happened was that the direction of 'confrontation' operations had passed to the Indonesian Army and Indonesian regulars were beginning to fight as units rather than as 'advisers' to half-trained gangs of terrorists.

This was discovered by the 2/10th Gurkhas in the Second Division of Sarawak in March. Sent to dislodge some Indonesians believed to be in position along the precipitous Kling Klang range of mountains that ran along the frontier, they ran into fierce opposition and were not surprised to find an enemy corpse dressed in the camouflaged smock and American equipment of the Indonesian Army. He belonged, it was discovered, to the 328th Raider Battalion, which proved itself well-trained in the tactics its name implied.

Only about forty Indonesians had defended their position but they had been difficult to dislodge and both sides suffered several killed and wounded. At the end of the month, the enemy were back on the ridge, two groups of forty being strongly established in caves opening on to a sheer cliff-face. The 2/10th called for artillery support and, when that proved insufficient, Walker asked permission to use the newly-arrived SS11 wire-guided missiles which could be mounted on helicopters. This was finally granted after reference to Admiral of the Fleet Lord Mountbatten, the Chief of the Defence Staff, in London. Wessex helicopters of 845 Squadron were armed with the missiles and flew straight at the cliff-face before launching their missiles into the caves. Again the Indonesians were dislodged, but this sort of fighting was not counter-guerrilla, it was war.

The campaign in Borneo was now being taken very seriously in London but not, it seemed, in the way Walker had hoped. While he wanted to prove to the Indonesians that not only could they not hope to win militarily but that they faced humiliating defeat if they persisted, the British Government appeared anxious to make it easy for them to change to a less aggressive policy without loss of face. A catch-phrase much used by British policy-makers overseas at this time was, 'Play it cool.'

Walker was appointed Commander of the Bath in the New Year Honours List but such recognition seemed overdue to many of his circle and had indeed been initiated earlier only to be delayed by what were described by those in authority as 'snags'.

Among Walker's visitors at the beginning of 1964 had been Mr Peter Thorneycroft, the Secretary of State for Defence, and Mr James Ramsden, the Secretary of State for War. Neither appeared to impress him as keen inquisitors, because while he insisted on briefing all visitors himself he liked them to ask searching questions, the more pertinent the better. But these two arrived in Brunei seeming to be satisfied that what they had been told in London and Singapore was all they needed to know. But, while he had them in his office, Walker took the opportunity to stress the success of unified command and that this had been established 'in the nick of time'.

A Labour Party politician who visited Borneo at this time greatly impressed Walker and the liking was mutual. This was Denis Healey, the Opposition spokesman on defence, who had made so deep and thorough a study of the subject that, even without access to official sources, he was regarded as a formidable authority. Healey accompanied Begg on one of his regular visits to Brunei and, as was customary, received a full briefing from Walker. Later the two men met for a private talk and Healey, taking out a notebook, said, 'Leaving aside the bullshit you gave me this morning, can we get down to the facts?'

'Those were the facts,' replied Walker. 'If you will accompany me on a tour of the frontier you can check them for yourself.'

They set off by helicopter, visiting the forward troops and Walker gave Healey every opportunity to question whoever he wished. Walker admired his guest's grasp of defence realities and the perception shown in the questions he asked and liked his blunt, direct manner. For his part, Healey was impressed by the intelligence and professionalism of the Director of Operations and made a mental note that when he came to office he would remember him.

Briefing visitors, a task which he never hurried, was one of many duties that were crowding in upon Walker. He was both Director of Borneo Operations (DOBOPS) and Commander British Forces Borneo (COMBRITBOR) which gave him tight, personal control over his command. While he controlled his air forces through his Air Commander and his sea forces through his Naval Commander, his control over land forces was direct. Brigadier Glennie, his Deputy, commanded Central Brigade and East Brigade, covering Brunei, part of the Fifth Division of Sarawak and Sabah; and West Brigade, which was responsible for the rest of Sarawak, was commanded by Brigadier F. C. Barton, of 3 Commando Brigade, and, from January, 1964, by Brigadier 'Pat' Patterson of 99 Brigade.

Thus Walker's detailed responsibilities were both upward to the Commander-in-Chief and GHQ in Singapore and downward through his brigade commanders, and they were both operational and administrative. A barrier and a filter was needed to ensure that only essential work came through to his desk and, in January, he chose a Military Assistant. His eye fell on Peter Myers, who had been his most trusted company commander in Burma, had fought through the twelve years in Malaya and who had already had experience of fighting in Borneo. As second-in-command of the 2/10th Gurkhas, Myers had arrived there early in the preceding year and had spent much time on detachment with two companies in the Third Division. He was now on leave and due to return to the Far East in the autumn to command the battalion. Walker now summoned him from London to join his inner staff, consisting of Glennie, his aide-de-camp and his personal assistant.

Myers' vitality matched Walker's, which was fortunate because his staff was expected to work a sixteen-hour day and longer when required. By the time Walker arrived in his office at eight o'clock in the morning, Myers would have gone through an average of sixty or seventy overnight signals and arranged them, tabulated in priority and subject on his desk. Immediately 'morning prayers', the daily staff meeting, would be held, the intelligence officer reporting on information that had come in during the past twenty-four hours. This would be followed by an hour or two of concentrated work in Walker's office then, unless there was an important visitor, there would be a flight up-country to visit units in the field, or an afternoon of paper work. When working in his headquarters, Walker would leave for Muara Lodge at half past five, taking more paper work with him and this would usually occupy him until well after midnight.

In the field, Walker was more concerned with asking questions than giving orders. These, issued later from his headquarters, would often be based on the answers he received. Always, he wanted to know not only what the present situation was, but how others thought it should be handled. Thus he made those he met feel personally and directly involved with the conduct of the campaign. Rebukes also would be delivered on these tours. With inefficiency and laziness, he was ruthless; with excusable lapses, he often employed pointed charm.

One such moment came when he visited the Argyll and Sutherland Highlanders and flew by helicopter to their forward positions with the second-in-command, Major Colin Mitchell, who was to become famous in command of the battalion three years later in Aden. As they strapped themselves into the helicopter, Walker noticed that Mitchell was wearing shorts. This was against the rules since long trousers were thought to give some protection against flash-burns in the event of an accident. But Walker chose not to administer a direct rebuke.

'Your knees are not very brown,' he said.

But Mitchell, understanding the purpose of the remark, did not rise to it. 'It is difficult to be sun-burned in the jungle, sir,' he replied with cool but polite cheek, since his job usually kept him out of the jungle at battalion headquarters.

Visitors who were entertained by the Walkers were surprised by the apparent normality of their home life. The guards, apart from the sentry on the gate and the soldier servants, were rarely seen. There were drinks before meals overlooking the sea, excellent food and wine and every comfort. The war seemed very far away.

Only once, in fact, was there a serious alarm. This was when the cook appeared in Walker's study one evening to report that a boatload of armed soldiers had been sighted approaching the beach. Walker ordered his ADC to call the guard and, since the sighting seemed genuine enough, called out the stand-by platoon. Telling Beryl and their daughter Venetia to get into the bath if shooting started – the bathroom had interior walls and was the safest place in the house – Walker poured himself a whisky and soda and went on with his work. Twenty minutes later the platoon with armoured car escort roared up to the house and patrols were sent out along the beach. It proved to be a false alarm: a boatload of fishermen dressed in green denims had been swept out to sea past Muara Lodge.

While his wife and daughter ran his household and Beryl was active in welfare work for the Services, the rest of the family became involved in the campaign. The twins had both passed through Sandhurst and Nigel, who had come to Borneo with the 1/6th Gurkhas, transferred to the Gurkha Parachute Company, and Anthony was later to arrive with the Rifle Brigade.

An occasional visitor to Muara Lodge was its owner, the Sultan, and Walker and he became firm friends. A small, slight man with a quiet manner, the Sultan was usually under-estimated by the British. Walker realised that he was not the covetous despot some tried to make out. The Sultan had spoken frankly about his problems and the difficulty in finding the right men for his administration. Some of his most promising young men had become involved in the rebellion, he once said sadly.

This friendship led to a generous gesture by the Sultan. Walker had been speaking of the poor living conditions for his troops in Brunei and, now that those on full-length tours of duty could be accompanied by their families, of the lack of suitable married quarters. The Sultan at once undertook to build new barracks and married quarters himself – at a cost of something over a million pounds – and to present them to the British as a gift. When news of this largesse reached Admiral Begg he was delighted, thanked the Sultan profusely and congratu-lated Walker on having been instrumental in bringing it about, adding, 'People have been knighted for less.'

But Whitehall was far from pleased. The Sultan, thought the civil servants, might be using this as an excuse to avoid paying his full share of the costs of suppressing the Brunei revolt. In any case it was none of Walker's business to usurp the duties of the Works Depart-ment. When Walker told the Sultan of this, he laughed, saying, 'I am not interested in the cost. Money cannot buy the gratitude I owe Brit-ain.' And Walker later remarked to a friend, 'On the one hand I was dealing with a sahib – on the other with a lot of Whitehall babus.'

Walker was constantly on the move. Apart from almost daily vis-its to his forces or to the headquarters in Kuching and Tawau, he flew to Singapore for regular meetings with Begg and to Kuala Lumpur for fortnightly meetings of the National Operations Council or the less frequent meetings of the National Defence Council of Malaysia.

Begg himself came over to Borneo every three months and would spend five days touring the theatre with Walker.

The burden of duties was heavy but, for one with Walker's energy, acceptable. What was more of a time-wasting irritation than necessary duty was the amount of administrative detail which, under Army regulations, he had to attend to himself. Walker had already challenged convention by down-grading specialist staff officers in the brigade headquarters so that instead of the service arms – the Royal Engineers, the Royal Electrical and Mechanical Engineers, the Royal Army Ordnance Corps – each having its own pyramidical organisation with, perhaps, a brigadier of their own at each summit, there was simply one efficient major from each arm to advise the brigade commander.

So Walker applied for a brigadier to take over some of his administrative work but was told that, as a major-general, he was not entitled to a staff officer of this rank and would have to do the work himself. So, in consultation with Begg, it was decided that Walker should relinquish the job of Commander British Forces Borneo but give it up only so far as it was concerned with administration. To find an officer ready to help circumvent the regulations in this way would not be easy. But another old friend came to the rescue. This was Peter Hunt, who was about to take over command of the 17th Gurkha Infantry Division and relieve Walker as Major-General, Brigade of Gurkhas. Most of Hunt's soldiers were either in Borneo, or resting between tours of duty there, so Walker suggested that he become COMBRITBOR in name, at least. But he made it plain that he had no intention of relaxing his own control of the ground forces and, although this would put Hunt in an invidious position, this would have to be accepted.

Hunt agreed and at the beginning of June, the three land, sea and air commanders were to set up their headquarters on Labuan leaving Walker and a small staff at Brunei until new quarters could be prepared for them on the island.

Because Hunt and Walker were friends the arrangement worked well, which might not have been the case since they were both major-generals. At the time, Walker agreed with Begg that the most sensible arrangement would be to promote him to lieutenant-general – temporarily and without extra pay if necessary – to enable him to have a staff suitable for so large a theatre of war. His superiors in the Army not only ignored the suggestion but, in Singapore and London, gossip circulated about Walker's personal ambition and his plans to inflate his command.

Such talk was not always malicious or envious. Despite the raids on Long Jawi and Kalabakan, the probing towards Kuching and reports of further Indonesian preparations in Kalimantan, there were still those who did not expect the fighting to escalate much further. Among these was Tom Harrisson, who had returned to the Sarawak Museum at Kuching, while maintaining a close liaison with the SAS headquarters camped in his garden.

Harrisson was cynical. His critics claimed that this was partly through pique at finding himself no longer one of very few authorities on Borneo, since British officers like Lieutenant-Colonel John Cross, who commanded the Border Scouts, had quickly learned languages and established happy relations with the tribes. Whether or not this was so, Harrison enjoyed propounding the theory that all soldiers escalated war because it was in their natural interest to do so. The many charms of Borneo intensified this tendency, he added, saying that a strong military presence had been established at his old wartime base of Bario, in the mountains south of Brunei, partly at least because the climate was pleasant, there were no mosquitoes and the natives were friendly.

Intelligence reports of Indonesian warlike preparations were, he claimed, sometimes exaggerated to this end. The long and tortuous lines of communication within Kalimantan alone dictated a low level of offensive action by the Indonesians.

A less sanguine view was taken by Walker and his staff and by Begg and his. If a CCO uprising coincided with massive border incursions, the situation would become desperate for the defenders. Even without major internal trouble, Walker was only able to defeat the Indonesian incursions so swiftly because they came singly, or only occasionally overlapped each other, and he was able to switch his forces rapidly by air. But were the Indonesians to co-ordinate their attacks he could be faced with an impossible task. A study of contingencies showed that, with the number of helicopters he had available, he could hardly hope to contain more than three simultaneous incursions on the present scale. Repeatedly he asked for more helicopters, only to be told that they were not available, or were needed by the Army in Germany.

The keystone of Indonesian strategy was, of course, that they held the initiative. They were on the offensive, they decided where to strike and when. Moreover they operated from safe sanctuaries, base camps which could be established along their border without fear that it would be violated by the defenders of Malaysia. From time to time, Border Scouts crossed into Kalimantan – as they always had – to visit villages belonging to their own tribe and brought back information. The SAS was, as usual, a law unto itself and on reconnaissance missions would sometimes penetrate into what was certainly enemy-held territory. But Walker's requests that he be allowed to force the enemy's base camps farther back into Kalimantan, or at least chase the enemy back over the unmarked border when in 'hot pursuit', were – although Begg fully sympathised with his arguments – firmly refused.

Between March and June, the new pattern had become apparent. In the Second Division a series of sharp actions were fought between Gurkhas and professional raiders, notably from the well-trained Black Cobra Battalion and, although the enemy withdrew with losses after each, this was a far more serious challenge than had been offered by the Indonesian Border Terrorists. In six battles fought in this area, the Indonesians lost a total of at least forty-five killed against three Gurkhas and they suffered two particularly humiliating, although minor, defeats. In one, four Black Cobras ambushed two Gurkhas but were themselves killed, their quarry escaping unscathed. In the other, six more were captured by Iban head-hunters and lost their heads, which, after being flown to Kuching for identification, were parachuted back to their new owners, arriving with the rations awaited by a Gurkha company.

June was a watershed point in the campaign; the beginning of an even more ominous phase. Inexorably the fighting had escalated. The collapse of the Brunei revolt had led on to the Tebedu raid in April, 1963, and this to the attack on Long Jawi in September. The massacre of the Malays at Kalabakan, at the end of the year, had deepened the conflict and, during the spring of 1964, Indonesian regular forces had become increasingly involved.

Now the negotiations between Indonesia and the Philippines and Malaysia, which had broken down in Bangkok in February, were resumed in Tokyo at the highest level. Presidents Sukarno and Macapagal met the Tunku for more fruitless talks, while a Thai mission visited Sarawak to witness the apparent withdrawal of Indonesian forces to Kalimantan. If proof were needed of Sukarno's cynical manipulation of Asian opinion, it was to be demonstrated now. The smart, well-equipped soldiers who were filmed and photographed marching out of Sarawak had, in fact, crossed over from Kalimantan, a little way along the border, earlier in the day. Nobody in Malaysia was surprised when, two days later, on 20th June, the Tokyo negotiations broke down and were abandoned.

Within twenty-four hours of the breakdown of the Tokyo talks, the Indonesians launched an attack that confirmed the change in tactics Walker had anticipated. At Rasau in the First Division, two platoons of the 1/6th Gurkhas, returning from patrol, camped in an old jungle base camp instead, as Walker would have expected, of sleeping under poncho capes. That night, about one hundred Indonesians assaulted the position and kept up their attacks for four hours – using mortars, grenades and automatic weapons – before withdrawing. It was not known whether they had suffered any casualties, but the Gurkhas had lost five killed and five wounded and this, in terms of such small-scale actions, amounted to a defeat. The Indonesians had shown professionalism and resolution and there now seemed no reason why the fighting should not only continue to escalate but do so at a mounting rate.

That month the new command structure in Borneo became effective, the subordinate Service commanders moving to Labuan, leaving Walker free to concentrate on the conduct of operations. At the same time the chain of command changed at the top. In the British General Election, the Labour Party emerged victorious and, to Walker's delight, Denis Healey was appointed Secretary of State for Defence. So, while his lack of confidence in some of his superiors within the Army remained, he was reassured that, while he remained answerable, through the Commander-in-Chief, to the Malaysian National Operations Committee, he also implemented the policies planned by two men he admired, Lord Mountbatten, the Chief of the Defence Staff, and now Denis Healey.

Walker's most urgent task at this time was to develop new tactics to meet the changing threat. His land forces were now in a full divisional organisation under General Hunt and their increasing experience was beginning to tell on the battlefield. All eight Gurkha battalions had now been committed and they were now returning after resting in Malaya for further six-month spells in Borneo. British troops spent only four months at a time in Borneo but they, too, were now beginning to return for their second tours. The Gurkhas had kept their jungle skills fresh but all British troops had to spend six weeks at the Jungle Warfare School in Johore before being committed to Borneo and, in Walker's view, did not begin to 'earn their keep' until they began their second tour. His experience of new arrivals was that any Far East defence policy based on quick reinforcement from Europe was doomed to failure. He also noted, sadly but without surprise, that the lessons learned with such labour during the twelve years of the Malayan Emergency, had largely been forgotten in two years, just as the lessons of the Burma campaign had been forgotten when trouble began in Malaya.

The Malayan experience would have been particularly important during the first year of the fighting in Borneo when operations were often similar. But, while there had been twenty-four battalions as well as gendarmerie and police to meet the internal threat in Malaya, he had had only half a dozen battalions and some auxiliaries to meet both internal and external threats in Borneo. But, in effect, he had been able to triple the strength of his 'teeth arms' by the skilful use of helicopters. These became so essential a part of all ground operations that they were regarded as being as much a part of military units in the jungle as Land-Rovers and armoured cars would be in a European theatre.

During this first year, Walker's tactics had been to establish a chain of patrol bases and helicopter landing zones along the length of the border. At first the bases were established in villages but, so as to minimise the chance of civilians coming under fire, they were later moved a short distance away, where they could offer protection without increasing the danger. A platoon operated from each of these bases, two-thirds of its strength being out on patrol at any one time, its own eyes and ears being supplemented by those of Border Scouts and SAS and other surveillance patrols. The density of this deployment varied with the threat but one battalion would often cover a front of over a hundred miles.

Each base had its own helicopter landing zone, or 'pad', but its own patrols sought the enemy on their feet. It was when they, or the small surveillance patrols, reported an incursion that the helicopters came into their own, bringing in the 'quick-reaction force'. During that first year, helicopter pads were cut and blasted in the jungle every thousand yards along the border so that, when the enemy was known to be at any particular point, troops could be landed within five hundred yards of that position.

When there was no pad available, soldiers slid down ropes from helicopters hovering over the leaf-canopy of trees that sometimes stood two hundred feet. These tactics were developed particularly by the two Fleet Air Arm squadrons, Nos. 845 and 846, under Walker's guidance. Nicknamed 'The Pirates' because of the colourful Iban dress they liked to wear off duty at their forward base at Nanga Gaat, they flew with extraordinary flair and dash. When a landing zone was cut in deep jungle, the news that it was 'wide enough for The Pirates' implied that its diameter was a few feet wider than that of a Wessex helicopter's rotor-blades. One such, into which Walker and Fenner were flown, was aptly known as 'The Funnel' and gave passengers the sensation of riding two hundred feet down a lift shaft.

Walker was adamant that helicopters should not be used in 'Prince Rupert tactics', headlong charges in formation such as were being rehearsed by the United States 1st Air Cavalry Division for use in Vietnam. Surprise was the most important element and helicopter pilots learned contour-flying, hugging the hillsides, skimming the treetops, slipping quickly into the landing zone. When engine noise made secrecy impossible, feint landings could be made in the hope of confusing the enemy.

Helicopters being relatively slow – however fast they seem at tree-top height – anti-aircraft fire was always a danger. This was particularly so on the frontier where the Indonesians sometimes mounted heavy machine-guns commanding known landing zones, and it was a gun firing across the border that shot down an Auster light aircraft at the end of 1963, killing a chaplain on his Christmas visits. Landing in the jungle was risky enough and troops on the ground ringed the landing zone, facing outwards. Take-off was more dangerous, as the arrival might have been seen and heard, and the first half-minute's flight was always a time of tension and acrobatic flying.

Essentially, Walker's tactics were now those of a boxer notably 'light on his toes' but, as it became clear that the Indonesians were capable of heavy and damaging blows, the defence had to be on a middleweight rather than a lightweight basis. Up until the spring of 1964 this had been a platoon commander's war, but then it became a war for company commanders. The old platoon patrol bases were replaced by company bases and the whole frontier began to assume a more warlike appearance.

Companies were now to operate from permanent, heavily-defended bases, or 'forts', and from these sallied out to establish temporary patrol bases in the jungle. There was no attempt at concealing the forts, indeed this would have been impossible. First, a chosen hilltop would be shaved of jungle and tunnelled like an ant-heap, deep trenches, or underground passages, connected dugouts with roofs of logs, corrugated iron and sandbags and firing positions giving all-round protection. These main defences were surrounded by barbed wire, minefields and trip-wires connected with mines or flares. A new weapon, which began to arrive, was the American-made Claymore mine, a small, flat, slightly concave container, standing on legs, which threw hundreds of sharp fragments forward in a concentrated arc.

In and around the wire, and particularly in long grass, was an ancient and terrible form of defence: panji stakes. These were lengths of bamboo sharpened to a point, so with no more pressure than that of a man walking slowly the spike would impale a foot or penetrate a thigh. These were particularly deadly because, unless treated immediately, the wound almost always turned septic, resulting in the loss of a foot, or even death. One of the most feared weapons of the campaign, panjis were issued in bundles to the defenders to plant round even temporary positions, and fine steel-mesh soles were fitted into jungle boots which the stakes would otherwise have penetrated.

These forts were armed not only with medium machine-guns and mortars in fixed positions but also with artillery. While fundamental artillery dogma is to concentrate guns, Walker now dispersed them so that one battery might be spread along nearly a hundred miles of frontier. Single 105-mm howitzers were slung under helicopters and flown to the forts, each gun's arc interlocking with that of another in the neighbouring company base. The most fastidious precision was required in the use of these weapons because the killing of civilians was in Walker's view unforgivable not only on humanitarian grounds but because he had come to rely so heavily on his 'hearts and minds' campaign.

From the beginning of the fighting in Borneo it was obvious that much of the weaponry and equipment, which was being issued from Singapore, was unsuitable for the terrain, the climate and the type of operations. Again, it seemed, the lessons of Burma and Malaya had been forgotten. The phrase that dogged Walker's requests for lighter, more durable warlike stores was 'general purpose', which implied that the particular item had been acquired for use in Europe and that other theatres must make the best of it.

None of the staple infantry weapons was satisfactory. The standard infantry rifle – the Belgian-designed, self-loading 7·62-mm FN rifle – was a good weapon for NATO purposes but, with its weight of eleven pounds and its long barrel, was too heavy and clumsy for jungle fighting, where action was likely to be at a range of between ten and fifty yards and very rarely at more than one hundred. Far more effective was the lightweight Armalite, deadly at short range and weighing about half as much as the NATO weapon. Its ammunition, which caused fearful wounds, also weighed less than half and this was a vital factor, particularly for long-range patrols which had to economise on weight to the maximum. Walker had studied tests carried out with it in Malaya in 1961, when its high-velocity capabilities had been tried on armour plate, sandbags and goats and it had undergone testing to the limits in the jungle. In response to Walker's urgent requests for this weapon and the realisation that the Indonesians were already armed with it, a consignment of one hundred was sent to Borneo for field trials and these were issued to selected battalions and generally ten were issued to each company on operations.

The Bren light machine-gun, which had come into service before the Second World War, was being withdrawn but it was replaced by the new general-purpose machine-gun which because of its greater weight was less satisfactory. There was a shortage of mortars because the old light, two-inch weapon, which Walker wanted, had been issued to the RAF Regiment for airfield defence and there was a shortage of ammunition for the new 81-mm mortar.

Much British equipment was too heavy and Walker had difficulty in convincing his suppliers that the urgency of equipping jungle forces on active service was at least as great as fitting out soldiers in NATO barracks in Germany. Wireless sets, built to last twenty years, were far too heavy and Walker wanted the American lightweight, long-range, six-channel sets that were designed to be thrown away and replaced rather than repaired. There was also an urgent need for new, lightweight, ground-to-air radio because the Army's wireless on the ground could not communicate with RAF aircraft overhead. Instead of green denims and under-pants, the Army needed zip-fastened lightweight suits, needing no underclothing, which were issued to the RAF in Borneo but not to soldiers in the jungle.

Weight was a constant problem in the jungle and had to be saved even at the cost of ammunition carried. Yet the ration packs were also 'general purpose' and included tins of food unsuitable in the heat and humidity. In vain Walker asked for rations planned specifically for jungle operations such as had been developed by the SAS and by the Australians.

Again and again Walker met some equipment problem that he had faced ten or more years ago in Malaya and thought he had solved. Sometimes the only knowledge that had then been gained, but had been retained, seemed to be in the fighting skill and jungle expertise of individual soldiers. One of his older officers remarked wearily that the infantry was now more unsuitably equipped than when it had started to fight the Japanese more than twenty years before.

It was particularly galling to note that the Indonesians were making the most of their experience and that some of this had been provided by the British. Until the formation of Malaysia had first been discussed, Indonesian officers had regularly attended courses at the Jungle Warfare School in Malaya where they had been taught the same tactics and tricks that were now being used against them.

Increasingly, the Indonesians practised what Walker had preached and, early in July, there was further escalation. The raiding forces stationed near the border inside Kalimantan, opposite the First Division, were reinforced; not to any great strength but enough to suggest that more or heavier attacks were planned. At the far end of the front there was increased Indonesian amphibious activity. The green berets of commando troops were seen, beach-landing exercises were reported and the Indonesians began jamming the Tawau Assault Group's wireless communications.

The internal threat continued, meanwhile. The CCO was now concentrating in spreading its influence from the towns into the country, particularly along the banks of rivers where Chinese planters and traders had settled. Also, in Sarawak, another subversive organisation, the Borneo Communist Party, had taken a more practical step in setting up its own little arms industry for the manufacture of shotguns and grenades. In Sabah, there were more signs of the activity of Indonesian agents among the immigrant workers and survivors of the Kalabakan raid, together with fresh infiltrators, were reported as trying to set up the framework of a guerrilla organisation.

These were reports that caught the eye but, along the whole border and to the north of it, at sea on either flank and in the air there continued a simmering on the edge of war that, in staff officers' jargon, is called 'low-intensity operations'.

In July, the Indonesians carried out thirty-four acts of aggression, according to the records kept at Walker's headquarters, and, of these, thirteen were border incursions. But, in August, this number fell by more than half, and for over a fortnight there was an almost complete lull. During these two months, the enemy lost about thirty-four killed in action against three of the defenders.

Walker had expected escalation; it had not continued, but he continued to expect it. Inevitably there were those, notably in Singapore, who again accused him of exaggeration and the chances of the weapons and equipment he needed being provided were sharply reduced. The security forces in East Malaysia and Brunei seemed well able to contain this level of threat and, it was pointed out, the jungle forts had been planned to resist odds of ten to one and the present attacks were nowhere near that level. Yet Walker's intelligence estimates were that the Indonesians were busy reorganising and planning and he remained convinced that this was no time for complacency, but rather for further warlike preparation.

This warning Walker gave to his first official visit from the new Labour Government in London. His visitor was Mr Fred Mulley, Healey's Deputy Secretary of State for Defence and Army Minister. In appearance and manner it would have been difficult to find two more different men in high office. The general: smart, precise, with a crisp air of command; the minister: a wartime sergeant and prisoner of war, small, untidy, with the benign manner of an old-fashioned Fabian Socialist. Yet the two men liked each other immediately. Walker was pleasantly surprised to find Mulley dedicated to his new job and quick to ask pertinent and intelligent questions. For his part, Mulley was impressed that Walker, despite his unmistakably military appearance, was totally without pomposity and obviously welcomed the most searching questions.

Among the points raised was the question of future strategy. There was a lull in enemy activity but there was no reason to suppose it would last – and what then? Were the Commonwealth forces to wait for the attacks, hoping that they had enough soldiers and helicopters to meet them? Or could they attempt to throw the enemy off balance before he attacked, so reducing the intensity of the fighting and even persuading the Indonesians that nothing was to be gained from aggression in Borneo?

The enemy held the initiative and would continue to do so while he could attack from secure bases in Kalimantan, but close to the border. These bases were safe from attack only because there had been no official declaration of war and the British Government was anxious to avoid taking any step that would be presented to the Afro-Asian and communist factions in the United Nations as 'imperialist aggression'. Under the present rules of the game, the Indonesians were free to violate the frontier but the Commonwealth allies were not. The frontier was unmarked and ill-defined – indeed, two versions of it appeared on some maps – and only by the most precise map-reading could the defenders stop themselves when in hot pursuit of retreating raiders. Surely, they could now be given permission to force the Indonesian bases farther back from the border? A few sharp lessons and the level of the threat would surely at once be reduced.

Although Walker had always held that attack was the best means of defence, his doctrine kept this within strict limits. There would be no question of launching a general offensive against Kalimantan but, rather, of keeping the enemy under pressure and off balance and, in doing so, giving him a clear idea of what could happen should he escalate the level of his own offensive. This view had the support of Begg and of all senior British officers directly involved – some, indeed, took the line that a sharp offensive by all three Services against certain strategic targets could decisively end the war – but seemed to face considerable opposition in Whitehall, including that of the Foreign Office and Field-Marshal Hull.

After hearing the arguments for such action, Mulley, it is understood, summed up the case against it. It was not for him to take a decision but to pass the opinions he had heard to Denis Healey, who could then ask for a decision at Cabinet level.

In August, the Indonesians raised the level of their confrontation, but not in a way that had been expected. This time they fell upon West Malaysia. In the early hours of the 17th, more than a hundred raiders

– three-quarters of them Indonesian marines and parachute troops and most of the remainder Malaysian Chinese communists – crossed the Straits of Malacca by boat and landed on the coast of south-west Johore. But instead of being greeted as liberators, as they had expected, they ran into Malaysian troops who had, in fact, been called from containing Chinese unrest in Singapore. Within a few days most of the raiders had been killed or captured and the survivors driven into the jungle, where they continued to be hunted.

A second attack, following a fortnight later, was airborne. Nearly two hundred parachute troops embarked in four transport aircraft at Jakarta, planning to drop near Labis, about a hundred miles north of Singapore, which lay on the railway line to central Malaya and had been a strongly communist area during the Emergency. Only three aircraft managed to take off, one of the others crashed into the sea, while trying to make its low-level run beneath the defending radar, and the two remaining ran into an electric storm over Labis and scattered their men over five miles of country.

This time the defenders were Gurkhas and none other than the formidable 1/10th, which had broken the raiders at Kalabakan nine months before. More veterans of Borneo quickly appeared on the scene when half of 845 Squadron, Fleet Air Arm, flew up from the *Bulwark*. Significantly, the Commonwealth became more deeply committed to defend Malaysia when the 1st Royal New Zealand Infantry Regiment arrived to join the hunt. Within a month, the raiders were virtually wiped out but others were to follow and it became clear that another dimension had been added to the undeclared war. In Borneo, Walker faced not only the internal and external threats but the possibility that raids on West Malaysia would deprive him of the reinforcements he might well need himself.

But, meanwhile, good news had arrived from London, via Singapore. The Government had approved limited cross-border operations up to a depth of five thousand yards inside Kalimantan. There was to be no public announcement, indeed the operations were to be carried out with maximum secrecy.

For several months, the SAS had, in fact, been crossing the border and penetrating enemy territory but only on reconnaissance. Four-man patrols had looked for the tracks of raiding parties and watched rivers, which were the main highways on both sides of the frontier. But the SAS had always worked in a mysterious way, never with other troops and answerable to other authority only at brigade level. Their patrols merged into the jungle for weeks and months on end and sometimes into Indonesian jungle for as much as ten days at a time.

Now, instead of watching and counting, they would be given permission to begin interdiction: ambushing tracks and rivers and setting 'booby-traps' where it was known that only raiders would pass. Sometimes their ambushes would be sophisticated affairs with electrically-detonated Claymore mines catching an enemy party front and rear while the SAS troopers' automatic fire raked the centre. Because the essence of SAS activity was speed of movement and reaction – they called themselves 'The Tiptoe Boys' – these were to be sudden, sharp little actions. The ambush parties hit and vanished, leaving the counter-attack to find nothing but apparently empty jungle.

The plan now was that infantry attacks could now be launched on a scale determined by circumstances to pre-empt any expected Indonesian attack. The front where such preventive operations were most urgently needed was in the First Division and Walker, Patterson, the commander of 99 Brigade, and their staffs had lengthy planning conferences. It was agreed that any temptation to wipe out enemy bases with the intention of inflicting as many casualties as possible – a temptation that was straining the self-control of several commanders of Gurkha battalions – must be resisted. Cross-border attacks on too heavy a scale would defeat their own purpose by forcing the Indonesians to retaliate in like manner, so escalating the war even further. Minimum force must be used with the object of keeping the enemy in a constant state of apprehension and uncertainty.

These operations were to be given the code-name of 'Claret' and were to be governed by what came to be known as 'The Golden Rules'.

Initially these guidelines were as follows:

Every operation to be authorised by DOBOPS – himself authorised by the Commander-in-Chief, Far East.

Only trained and tested troops to be used. No soldiers were to go across during their first tour of duty in Borneo.

Depth of penetration must be limited and attacks must only be made to thwart offensive action by the enemy and must never be in retribution or solely to inflict casualties. Civilian lives must not be risked.

No operation which required close air support – except in an extreme emergency – must be undertaken.

Every operation must be planned with the aid of a sand-table and thoroughly rehearsed for at least two weeks.

Each operation to be planned and executed with maximum security. Every man taking part must be sworn to secrecy, full cover plans must be made and the operations to be given code-names and never discussed in detail on telephone or radio. Identity discs must be left behind before departure and no traces – such as cartridge cases, paper, ration packs, etc. – must be left in Kalimantan.

On no account must any soldier taking part be captured by the enemy – alive or dead.

The first troops to be used for 'Claret' operations would be Gurkhas, it was decided, because they were the most experienced in jungle fighting. Initially, at least, they would operate on a single company basis for this was the scale of operation to which they were accustomed. There would, for the time being, be no more than one incursion made on each battalion front at any one time and these would be tightly controlled, the reins stretched back to Walker's own hands.

Already a certain vagueness had descended over official announcements about the campaign. For diplomatic reasons, all units of every nationality were combined under the title of 'security forces', this including everybody from the SAS to local police and Border Scouts. Now the smoke-screen was to become opaque.

The first 'Claret' operations were virtually extensions of routine ambushing but carried out a mile or two farther south under conditions of extreme secrecy. It is doubtful whether the Indonesians realised that they were seeing the beginnings of a new Commonwealth strategy. This was partly because the new series of actions took place so near to the frontier which was, in any case, so loosely defined and partly because their own communications and staff structure were far less able to cope with a flow of reports and in assessing them than were Walker's. While Patterson could fly up to his forward positions in the First Division by helicopter in fifteen minutes, Indonesian commanders, a hundred miles from the same front at Pontianak, had a flight of at least an hour in one of their comparatively few, serviceable light aircraft or helicopters.

These early raids achieved the same sort of limited success as did the routine ambush north of the border. In one, forty Indonesians were ambushed on a jungle track opposite the First Division and six killed, four wounded. While keeping to within five thousand yards of the border, there was little change in the Gurkhas' tactics, or in their routine, with the notable exception of security. The need for secrecy had been so forcefully impressed upon those taking part that not even the officers of a battalion involved in 'Claret' spoke of it in the mess as they had freely discussed earlier operations. This self-restraint, combined with the sinister routine of leaving behind all marks of personal identification – letters from home and family photographs as well as identity discs – and the knowledge that the enemy had been known to torture prisoners, combined to produce a climate of apprehension.

Officers of Gurkha battalions, which had been returning to Borneo for six-month tours for nearly two years, had been under considerable strain, for stress could result from prolonged jungle patrolling even without making contact with the enemy.

Thus, for the first time, there was a morale problem in Borneo. It came nowhere near the morale crisis that was to affect American troops in Vietnam, nor did it affect efficiency, but, in several battalions, there was an unaccustomed, unexpressed feeling of dread. This did not last. Realising that lack of communication lowers morale, security within the battalions was relaxed and, once the officers were free to talk about their expeditions over the border among themselves, morale rose to its former high level.

'Claret' raids were only planned and, after about a fortnight's preparation, launched, when there was evidence of a suitable objective. Thus, because the number of incursions threatened by the Indonesians during the final months of 1964 was small – ten actually launched in October, five in November, four in December – the number and strength of counter-raids was equally small and did not significantly escalate the level of fighting. As a result, Walker was again accused of 'crying "Wolf!"' Those who believed the campaign to be one of internal security, those who thought that the Indonesians could not and would not stage a major offensive and those who claimed that Walker and his Service commanders were exaggerating the threats for their own power and glory, were joined by the diplomats and senior officers in Whitehall who feared that counter-aggression would raise rather than lower the level of crisis.

Walker himself was convinced that his forecasts were well-founded and, in December, he was shown to be right. At the beginning of the month, soon after Patterson handed over command of West Brigade, covering the First, Second and Third Divisions, to Brigadier W. W. ('Bill') Cheyne, and Brigadier Dato Ismail handed over his West Brigade to Brigadier Tunku Burhanuddin, changes were noted across the border in both areas.

Opposite the First Division, Indonesian troops, who had become familiar over the past months, were withdrawn and fresh formations took their place. These in turn were reinforced by what appeared to be at least one complete infantry brigade. Opposite Tawau, enemy reinforcements began arriving and it was estimated that these would bring the regular KKO marines, who supported the irregular border guerrillas, up to brigade strength. And there were intelligence reports of even stronger Indonesian formations moving forward.

Before this reinforcement began in mid-December, Indonesian offensive forces along the frontier had been a mixture of regular companies, each up to two hundred strong, and volunteer companies of irregular guerrillas. Opposite West Brigade there had been more than eight regular companies and at least eleven irregular. Opposite Central Brigade, defending the Fourth and Fifth Divisions, Brunei and part of Sabah, there had been more than six regular companies and three irregular. Opposite East Brigade, concentrated on the coast around Tawau, there had been four or five companies of regular marines and about three companies of irregulars. This force had a total strength of twelve thousand men.

Observation by patrols, Border Scouts and intelligence agents now showed that the enemy was about to double his strength. More than fifty companies of regular infantry or marines – three complete brigades in the west and one in the east – were in the process of moving forward to within twenty miles of the frontier and there might be more to follow. By February it was expected that there would be well over twenty-two thousand soldiers in about fifty regular companies and twenty irregular companies along the border. While the marine brigade and its irregulars would face Tawau in the east and the central front was likely to be thinly manned by only one battalion, the bulk of this army would be concentrated opposite the First and Second Divisions, making a direct attack upon Kuching not only possible but probable.

This build-up was being planned and carried out by the Commandant of the Inter-Regional Command of Kalimantan, General Maraden

Panggabean. An experienced fighting soldier aged forty-two, Panggabean had been trained first by the Japanese, then in the United States and had completed his military education at the Indonesian Command and Staff College. Trusted by Sukarno, he had specialised in putting down attempted rebellions against the regime, in doing so rising from battalion to divisional commander. His last appointment had been Chief of Staff to the Inter-Regional Command for East Indonesia, responsible for planning the expulsion of the Dutch from West Irian.

Under Panggabean's command, the fighting forces were being regrouped as No. 4 Combat Command and their Director of Operations was to be a Colonel Supargo. He was aged thirty-nine, had been a soldier for twenty-two years and was obviously an officer of high quality. Supargo had been a brigade-major before being sent on a course with the United States Army at Fort Benning and serving as Assistant Military Attaché in Washington from 1954 to 1956. Significantly, he had been appointed Military Attaché in Kuala Lumpur and, as such, had been able to study the curriculum of the Jungle Warfare School, through which some of the best officers now under his command had passed. Finally and ironically, he had, less than two years before facing Walker across the Borneo jungle, passed out from the Pakistan Army's Staff College – the old Indian Army Staff College at Quetta – twenty-one years after his adversary.

But the total force available to Supargo was, of course, far greater. Within Sarawak, there were now estimated to be two thousand 'hard core' CCO terrorists together with twenty-four thousand Chinese sympathisers and another three thousand Indonesian immigrants. Within Brunei, there remained such widespread unrest that the withdrawal of British forces would, it was believed, lead to another rebellion which would depose the Sultan, this time to set up a communist state in the image of Cuba. In Sabah, there remained the potentially explosive immigrant Indonesian population of twenty thousand, more than half of them in the Tawau area.

Thus, if the worst came to the worst, Walker could find himself facing a combined total of something like seventy-five thousand potential enemies, ranging from well-trained Indonesian commando troops
armed with the most modern weapons to the simple terrorist ready to
throw a grenade. It was, Walker and Begg agreed, an ominous beginning for 1965.

What were Indonesian intentions? What were the most threatening dangers?

A rising of the CCO and the Indonesian immigrants coinciding
with an offensive by three Indonesian brigades with the objective of
capturing Kuching remained the worst possibility. There were no natural barriers to be defended between the border and the capital of Sarawak and the security forces might be unable to hold it. Eventually,
it was agreed, Commonwealth forces would retake the city and drive
the Indonesians back but, by that time, the situation might be out of
hand, with guerrillas in virtual control of the country and rebellion in
Brunei.

Nor was it likely that the war could be confined to Borneo. The
Indonesians had kept up their attacks on West Malaysia and sea-
borne landings – several in company strength – continued and, while
they were still unsuccessful, it was a short escalation from this to open
war. Indonesian warships were active in the Straits of Malacca, saboteurs were being put ashore and fighting at sea also seemed both
possible and probable. British contingency plans had already been
put into action and strategic naval and air forces were in the Far East
in readiness for open war with Indonesia.

To meet these threats, the forces available to Walker in Borneo
were pathetically small. West Brigade, with a front of six hundred and
twenty-three miles, now had five battalions – one British, three Gurkha and one Malaysian – and twenty-five troop-carrying helicopters.
Central Brigade, with a front of two hundred and sixty-seven miles,
had two Gurkha battalions and twelve helicopters. East Brigade, with
a front of eighty-one miles, had one commando and one battalion

forward and one Malaysian battalion defending in depth and no troop-carrying helicopters. The total number of soldiers available to Walker in Borneo was a little more than ten thousand.

Reinforcements were now a matter of urgency but it sometimes seemed to Walker's staff that Whitehall would never understand the need. Requests for a few staff officers or specialist troops were still rebuffed with the excuse that they were required in Germany to support NATO. This attitude had again been demonstrated when Walker was visited by General Hewetson's successor in Singapore. This was Lieutenant-General Alan Jolly, who had arrived from Whitehall where he had been Vice-Quartermaster-General, an appropriate stepping-stone to the new appointment which would be much – if not mostly – concerned with providing Walker's soldiers with weapons and equipment.

Shortly before Jolly had arrived in June, the command structure in the Far East had been reclassified to fall into line with the new Ministry of Defence, in which all three Services and the old policy-making Ministry of Defence had been combined in April. The titles of the three Service Commanders-in-Chief were to be replaced by the more modest appointment of Commander and it was as Commander Far East Land Forces that Jolly came to Singapore. A tank specialist, he had fought from Normandy into Germany in 1944 and had acquired a reputation as a painstaking professional and a pleasant person.

Although Walker and Jolly were as different as two men could be, and the latter had heard of the former's reputation as an uncompromising zealot, he arrived in Brunei determined to establish an easy relationship. But Jolly's background and manner were against him. To Walker it seemed absurd to send an authority on armoured warfare, who had commanded an armoured division in Germany, to a position of high responsibility in the realms of jungle warfare. 'I am a light infantry chap,' Walker told a friend. 'I am used to carrying seventy pounds on my back and not making a noise. Jolly is used to carrying seventy tons on his back and making a noise.' He also felt that Jolly lacked the authoritative presence – which he and Begg had in abundance – that was admired by their Asian allies. This first meeting between the two men demonstrated both the attitude that Walker felt he had to adopt in the face of scepticism in Whitehall and the difficulties his own seniors sometimes had in establishing relaxed relations.

Walker began by telling Jolly of his need for reinforcements. At this time the massive Indonesian build-up had not begun, so, asked to specify his needs, Walker mentioned the Royal Engineers. Sappers were needed both for operations – to build airstrips, roads, bridges and so on – and for the 'hearts and minds' campaign – to build schools and medical centres and sink wells, for example. Jolly replied that Sappers were not available and could only be provided at the expense of the British Army of the Rhine. Walker said that he was not there to argue about the need for Sappers. He was just letting him know what the requirement was and that he was about to state his case and make an urgent request for more Sappers to the Commander-in-Chief, Far East.

Jolly pointed out that some people would consider the need for Sappers in Germany to be more pressing than in Borneo. That, replied Walker, was a matter for Whitehall to decide when his urgent request had been made. He was telling Jolly of his intentions purely through courtesy, for there was no need for him to do so.

Next morning, Jolly said to Walker that he had been thinking over his case for Sapper reinforcements and had decided to support a request to the C-in-C. 'After all, I am his Service adviser,' said Jolly. 'And I am his Director of Operations,' said Walker.

One source of reinforcement was always available. This was from Malaysia and, by the summer of 1964, three battalions of the Royal Malay Regiment were serving in Borneo and more, training in West Malaysia, would follow. Since the debacle at Kalabakan, the Malaysian troops had fought well on several occasions but they could hardly be expected to achieve the standards set by seasoned Gurkha battalions. They were also facing some of the problems faced by most armies when sending inexperienced troops on active service. The Malays were proud and smart and they not unnaturally expected some of the soldier's traditional rewards. The young officers – like their British counterparts – hankered after the limited, but lively, night life of Kuching and their men after the comely Bornean girls in the longhouses. Probably the most important rule in Walker's 'hearts and minds' campaign was the avoidance at all costs of any involvement with Bornean women. One scandal and an entire longhouse, or even a whole tribe, might swing its sympathies towards the Indonesians. Tom Harrisson had written in one of several booklets on etiquette for soldiers in Borneo, 'If you have to pinch someone's bottom – count ten and then pinch your own.' In Malay battalions this warning was not always heeded.

From the beginning, the Malaysians had insisted that their soldiers be given positions of danger and honour and this had been done. They had gone to Tawau in East Brigade, commanded by a Malaysian officer, and to the most crucial front of all in the First Division. Here Patterson had to balance the importance of making use of the Malaysians' military aspirations and of prudence in exposing inexperienced troops to too much danger. Malaysian martial pride could be hurt just as easily by being defeated in action as by being kept back in comparative safety.

Patterson's dilemma over the Malaysians was seen by the author of this book earlier in that year. Arriving to visit the brigadier at 99 Brigade Headquarters in Kuching, Patterson announced that there had been a shooting incident on the frontier involving the Royal Malay Regiment, that he was flying there at once and extended an invitation to accompany him.

Flying south by RAF Whirlwind helicopter, Patterson stopped at the Malays' battalion headquarters to collect their commanding officer and to hear further details. Apparently a platoon patrol base at Nibong, on the frontier, had come under automatic fire and an attack was expected. Flying on, the helicopter descended to treetop height, zigzagging to avoid possible anti-aircraft fire, as the landing zone at Nibong was within easy range of machine-guns inside Kalimantan.

As the Whirlwind turned steeply and slipped into the prepared clearing, Malay soldiers could be seen deployed around the edge of the landing zone but, instead of looking outwards into the jungle, they were staring at the helicopter to see who were the passengers. Patterson, immaculate in tropical dress, red-banded cap and regimental cane, got out and asked the Malay lieutenant commanding the post what had happened. Guiding him to the sandbagged sangars, where Malay soldiers crouched behind their weapons, the young officer described how, a few hours earlier, they had been fired upon. Patterson asked two questions. Had they identified the enemy's firing position by finding spent cartridge cases? Had they sent out a patrol to cut off the enemy from the frontier? They had not.

Then, to everybody's alarm, Patterson strolled over to the edge of the jungle, where the shooting was said to have taken place, and began brushing grass and leaves aside with his cane in a search for the cartridge cases. As he did so, he gave the Malay lieutenant some advice. If there was further firing he must at once send out a patrol to get behind the enemy and ambush him. This was, of course, the reflex action of any attacked patrol base but, in the excitement of the shooting, the young officer had forgotten. But he would learn.

Patterson then walked slowly over to his helicopter – this, too, within the field of fire of any enemy still lurking at the jungle's edge – and took his time over the farewells before the fast, skidding take-off. It would be unlikely that this particular platoon would forget this lesson, for there was no question about their enthusiasm and pride; but they should not have to learn such basic lessons on the battlefield itself.

So, when cross-border operations were being planned, only SAS and Gurkha soldiers were to be used, and there could, as yet, be no question of involving the Royal Malay Regiment.

This was a practical decision but, already, General Tunku Osman, the Malaysian Armed Forces Chief of Staff, was asking when his soldiers were going to take the offensive. He not only asked this in West Malaysia but came to Kuching with the intention of putting it to Patterson himself. Like Walker after the Kalabakan disaster, Patterson felt certain that complete frankness between allies was essential. He knew and liked General Tunku Osman and resolved to tell him, as one soldier to another, that the Royal Malay Regiment was not yet trained and seasoned to the point where it could be used in dangerous operations.

He spoke in these terms, but the Tunku Osman reacted violently. He reacted just as Fenner feared his Prime Minister would if Walker had spoken frankly about Kalabakan. He was outraged, furious and almost in tears. He refused to believe that Malaysian soldiers were not the equal of any in the world. Moreover, he seemed to take Patterson's frankness as a personal insult.

On his return to Kuala Lumpur, General Osman sent an enraged complaint to Begg, demanding not only an apology but that Patterson be replaced. Begg at once asked for an explanation from Walker, who flew to Kuching to seek one from Patterson. It was obvious to the British officers that Patterson had spoken in friendly frankness but now equally obvious that, at this stage of national development, Malaysian pride was very sensitive indeed. The affair was brought to a close with tact and diplomacy. Even so, for several months to come, the Chief of the Malaysian Armed Forces Staff refused to receive Brigadier Patterson, until after he was relieved, in the normal course of events, in December by Brigadier Cheyne.

The reinforcements Walker now needed were far more than a few specialist supporting units and newly-raised Malay battalions. To face massive internal and external threats, he needed a strong reinforcement of the best infantry battalions available.

Yet he stressed that his need was not so much of large numbers of soldiers but of more helicopters which could, in effect, triple the strength of his ground forces. At that time he had about forty troop-carrying helicopters but, of these, at least eight were out of action for technical reasons. Now, to give each deployed battalion six helicopters, another thirty-six were required.

The Chiefs of Staff Committee agreed that reinforcements were necessary and should be despatched. Two more infantry battalions – bringing Walker's total up to twelve – would be sent, but only another twelve Whirlwind helicopters. Walker protested that the new battalions, together with another from the Royal Malay Regiment that was to be redeployed in the East Brigade area, would also need their own helicopter lift: another eighteen aircraft. Thus, he urged, his minimum requirement was for another fifty-four troop-carrying helicopters.

Meanwhile the structure of the Borneo forces had been changed. During the autumn, Walker had moved his headquarters to Labuan, where the three Service commanders were installed with their staffs and were shortly to be joined by the headquarters of the SAS. This improved communication and communications: personal communication between the commanders and their staffs and communications with the key centres of Kuala Lumpur and Singapore, Kuching and Tawau. It also meant that Walker could survey the whole theatre from a more detached viewpoint, although he retained his quarters at Muara Lodge outside Brunei Town, from which he commuted daily by helicopter.

So that he could concentrate even more fully on the direction of operations, and also to placate the frustrations of the three Service commanders in Singapore, these were now to take command of their own Services in Borneo for all but operational purposes. But by now they well understood that any interference with operations would bring the wrath of Walker – and probably Begg – upon them.

The two reinforcement battalions were to be deployed in the First Division. However, the strength of West Brigade would remain at five battalions and the added strength used to form another brigade to take over the Second and Third Divisions. Under this arrangement, West Brigade's front would be reduced from six hundred and twenty-three miles to one hundred and eighty-one and the new Mid-West Brigade, with headquarters at Sibu, commanding the Royal Malay Regiment battalion from the First Division and a Gurkha battalion, would hold a less-dangerous front of four hundred and forty-two miles. The responsibilities of Central and East Brigades were unchanged.

In January 1965, a thirteenth battalion – the 1st Gordon Highlanders – was sent to Borneo from Britain in the first direct reinforcement. They went on a jungle training course in Sabah before being committed on the East Brigade front.

Even with his reinforcements, Walker's forces seemed modest in the face of the threats. He would now have a little more than fourteen thousand soldiers – including five batteries of artillery (twenty-nine guns), two squadrons of armoured cars and four field squadrons of engineers – and less than sixty troop-carrying helicopters to cover a front of a thousand miles and a threatened interior the size of England and Scotland. In support there were, throughout the area, the equivalent of two battalions of the Police Field Force and about one thousand five hundred Border Scouts. The need for preventive, cross-border operations would not be reduced by the reinforcement.

In London, the Labour Government now showed its trust in Walker and Begg by not only allowing 'Claret' raids to continue but to increase their scope. By stages, the depth of penetration was increased from five thousand yards to ten and then to twenty and permission was given for small-scale amphibious raids to be made on the coast round either flank, particularly by the Royal Marines' Special Boat Sections. Inland, the strength of the raids was stepped up until a whole battalion might be involved, with two of its companies over the border. As British infantry became more accustomed to jungle operations, they joined the Gurkhas in taking the offensive.

The principal objective of these raids was to throw impending raids off-balance. When, at times, Walker considered that enemy casualties had been too heavy he would relax the pressure in that sector, even when a battalion commander reported that he was in a position to wipe out an entire Indonesian company. Whenever possible, consecutive raids would take place far apart – sometimes at opposite ends of the frontier – so that the Indonesians, with their poor lateral communications, would be less likely to relate them to a general strategy.

Walker kept his raiders on a tight rein, insisting that his 'Golden Rules' be strictly obeyed and, if it was thought essential to break them, this could only be sanctioned by himself. One such moment came when an urgent request was received from a Gurkha raiding party inside Kalimantan for helicopters to evacuate their wounded.

Up to that point, the raid had been successful. Crossing the border from Sabah, the Gurkhas had ambushed the enemy and inflicted fifteen to twenty casualties. But, in this action, two of their riflemen had been seriously wounded, had lost much blood and seemed certain to die if not evacuated immediately by air, as would have been automatic had they been hit on their own side of the border. The battalion commander, controlling the operation from Sabah, made an urgent request for helicopter evacuation. This was passed from Central Brigade at Brunei to RAF headquarters on Labuan, where it was refused by the Air Commander. It could now only be granted by the Director of Operations himself.

Walker spoke to the battalion commander by wireless and was told that the two Gurkhas would certainly die if not flown out. How long would a helicopter have to be over the border? asked Walker. Not more than twelve minutes, he was told: five minutes flying in, five minutes flying out and two minutes on the ground. Could it also be guaranteed that the helicopter's landing zone in Kalimantan would not be within range of Indonesian small-firms fire? He was told that it could. Then Walker gave him permission to attempt the rescue.

It was a calculated risk that had to be taken, but not only on humanitarian grounds. Sometimes soldiers' lives had to be lost for purely tactical or political reasons. But if these two Gurkhas were left to die then morale, which had recovered from the early apprehensions over cross-border operations, might suffer, for soldiers north of the border had always been comforted by the thought that if wounded they would be flown to hospital, probably within the hour. Now, it might seem that a light wound over the border would mean death. But, should the helicopter be shot down, or crash, then Sukarno could present its wreckage, and its dead or captured crew and Gurkhas, as proof of aggression against Indonesia.

Walker waited for another report. It came within half an hour and was that a Wessex helicopter of 845 Squadron, Fleet Air Arm, had flown into Indonesia and brought back the two Gurkhas. 'You have saved their lives,' he was told.

The most urgent need of 'Claret' operations was in the First Division to hold off the threat to Kuching. Here one of the most successful incursions was launched, one that was to set a pattern and standards for future attacks, by a Gurkha battalion.

The first raid by a company had the ambushing of river traffic as its objective. After two weeks' exercising in Sarawak, the company was 'let off the hook' and crossed the border undetected by the Indonesians. It had been decided that, if the targets seemed promising, the whole company should be used in a single ambush and it took up positions along the bank of a deep river known to be used by the enemy for ferrying troops up to the frontier.

The river was swarming with boats, but fire was held until a suitable target arrived. This appeared in the form of a landing-craft loaded with about thirty Indonesian soldiers. As it entered the planned 'killing ground', the order to fire was given and machine-guns virtually shot it to pieces, killing all on board. After such a success it would have been usual to withdraw as quickly as possible to the safety of Sarawak. But, although mortar bombs were lobbing on to his position from the far bank, the company commander decided to stay where he was in the expectation that the Indonesians would imagine that he had withdrawn.

Five hours later, a motor-boat with another boat in tow, manned by nine Indonesian soldiers, appeared on the river. They, too, were killed instantly. Even then the Gurkhas did not withdraw, waiting until nightfall and returning over the border under cover of darkness.

The same company made another raid three weeks later into a part of Kalimantan that had not hitherto been penetrated. To save time, the Gurkhas moved down a jungle track by night but, at first light, were seen by an Iban rice-grower and, expecting to be reported to the Indonesians, left the trail for swamp and dense undergrowth. There they clipped a large shelter in the thick foliage, established observation posts around a small perimeter and sent out patrols.

As was feared, the Indonesians had been alerted and a patrol, ten strong, came down a track nearby, obviously in search of them. Two Gurkhas, hidden in their observation post, watched the Indonesians approach, then sit down two yards away from them and cook a meal. There they stayed for five hours, eating and talking, then rose, buckled on their equipment and disappeared down the track.

The company commander waited until night, then made contact with his headquarters in Sarawak by his 'whisper transmitter' wireless, reported that the Indonesians had been searching for him and asked permission to ambush the jungle tracks in the hope of catching his own hunters. At three o'clock in the morning, the company clipped its way out of the hide and, at first light, three platoons were in ambush positions on the network of tracks.

The first ambush was sprung seven hours later. Ten Indonesian soldiers came down the track towards a killing ground covered by a machine-gun manned by two Gurkhas. Unsuspecting, they walked into it and a whole belt of machine-gun fire, felling all ten. But only nine were dead and, as the Gurkha machine-gunners moved to check the results of their shooting, the tenth Indonesian, who had been feigning death, shot and killed them both.

More firing now broke out and two more Gurkhas were killed and two wounded. The engaged ambush parties withdrew from their positions but one section commander stayed where he was, hoping to kill the Indonesian who had shot the machine-gunners. Stalking each other, the two men sprang to their feet five yards apart, fired a magazine each and missed. It was the Gurkha lance-corporal who had a reserve weapon, fired again and killed his resolute enemy.

Artillery fire from Sarawak was called down to cover the evacuation of the dead and wounded from the battle area but the company remained in Kalimantan for another day, withdrawing seven thousand yards, and across three rivers, to safety.

As each 'Claret' raid gave Walker's soldiers increased confidence, so these became more routine and soon there were more actions to the south of the border than to the north. So meticulously were these planned and rehearsed that they were usually successful and, even when the intention was to kill only two or three Indonesians in order to keep the rest in a state of apprehension, their casualties were often much higher. Full-scale raids were always planned by the headquarters of the Director of Operations and the brigade and battalion concerned and finally authorised by Begg in Singapore. But so commonplace did they become that brigade commanders were allowed to send over their own reconnaissance patrols up to the limit of twenty thousand yards on their own initiative, provided they kept Walker informed. He himself kept his forces under such tight control that he carried the positions not only of his raiding parties but of his surveillance and reconnaissance patrols and intelligence agents in his head. A criticism of Walker at this time was that he held the reins almost too tight.

By the beginning of 1965, Walker was confident that he had the measure of his opponent, Colonel Supargo. While he respected the Indonesian regular forces' qualities of professionalism and courage he knew that, fighting on equal terms as he now could, the high standards of training he insisted upon would prevail. Above all, the flexible tactics, dependent upon fast movement of soldiers by helicopter, which he had developed, gave him the confidence of a master-duellist, standing back from his opponent, parrying and lunging at will. While he could launch his 'Claret' raids, the enemy could not brace himself for a decisive attack.

Walker was more than content that he had been able to achieve this superiority at this time for personal, as well as military, reasons. In the spring of 1965 his appointment was coming to an end and Admiral Begg, for whom he had so much admiration, was himself leaving the Far East a few months later to become Commander-in-Chief, Portsmouth, and of the NATO Channel Command. Walker had been told that his own next appointment would be a very different one and had been decided upon in order to fit him for future, greater responsibilities in Europe. He was to be given one of the key staff jobs in NATO, Deputy Chief of Staff (Operations and Intelligence) at Headquarters, Allied Land Forces, Central Europe, where he would be directly responsible for facing the threat of a Russian attack on the West through Germany towards France and Britain. His active experience in Europe had been limited to the short spell with Eastern Command in England so, if he was to expect promotion in rank and position, then an appointment in NATO was essential.

His relief in Borneo was to be Major-General George Lea, an infantry officer, whom he had known in India just before the Second World War, and who had won a high reputation while commanding the Special Air Service. Walker was confident that he would hand over efficient land, sea and air forces, trained in the new concept of fighting that he had developed in the two years he had been fighting in Borneo.

But he was not content with the weapons and equipment he had been given for his campaign and feared that this sluggish support from the Ministry of Defence could be a symptom of a reluctance to face the underlying dangers of the Far East crisis. Reading reports from Vietnam of heavy defeats suffered by the South Vietnamese army at the hands of the Vietcong guerrillas, despite American air support, and the likelihood that the United States would have to commit ground forces, he could foresee similar possibilities in East and West Malaysia. The Americans were failing to hold the Communist offensive in Vietnam, but, in Borneo, he had stopped his enemy in his tracks. But could Whitehall appreciate the fearful possibilities?

Now, at last, it seemed that the Government was showing suitable interest in the undeclared war it had been fighting since 1962. Admiral Mountbatten was himself coming to Borneo early in the New Year, bringing with him Sir Solly Zuckerman, the Government's Chief Scientific Adviser. They were due to arrive in February – and to be followed by a visit from the Duke of Edinburgh – so Begg suggested that Walker postpone his departure until after their tours.

As was his custom, Walker prepared to deliver the briefings to his visitors himself. This time he not only took extra trouble in preparing maps and charts but drew up a list of his urgent requirements in the knowledge that he could now make them directly to the highest authority instead of seeing them disappear into a bureaucratic machine, usually without trace.

Admiral Begg met Mountbatten and Zuckerman at Singapore and accompanied them to Kuching, where they were met by Walker. During the briefings that followed, Mountbatten showed particular interest in the success of the unified command system, which he had inspired. To both his visitors, Walker stressed the need for suitable weapons and equipment. Some of this had by now been promised but its delivery was to be in small quantities and usually at some unspecified date several months ahead.

Walker listed his particular requirements. Among them were lightweight ground-to-air radio, ration packs and sleeping bags. He stressed the need for a new jungle boot to replace those being issued, which sometimes fell apart after ten days of hard wear. But, above all, he pointed out the urgent need for the Armalite rifle. Battalions which had been armed with the sample supply of one hundred had found the weapon an unqualified success. He had asked for another two hundred but these had not yet been forthcoming. He now asked that the Armalite become the standard jungle-fighting weapon, ordered in thousands.

It was fortunate that when Walker took Mountbatten and Zuckerman to visit a battalion that had been using the Armalite, it was that commanded by Peter Myers and had also just come out of action in which several enemy had been killed with this weapon. Talking to General Hunt, the Borneo Land Forces Commander, Myers and the soldiers who had handled the Armalite, convinced them that it was all that Walker had claimed. What his pleading had failed to achieve over many months was instantly accomplished. In a signal to London, Mountbatten said that he was satisfied that the Armalite trials had proved conclusively that it was a firm requirement for jungle operations. He urged not only that an order be placed but that it should be made immediately to pre-empt an expected order for eighty-five thousand Armalites from the United States Army for use in Vietnam.

As result of the Mountbatten visit a change came over the attitudes to the Borneo campaign apparent in London. At last it was no longer an embarrassing military sideshow but a confrontation and conflict of the utmost importance. The developing horrors of Vietnam showed all too clearly what could happen if limited aggression and subversion were allowed to succeed. To Walker – and probably to Begg and Mountbatten – it demonstrated the advantages of unified command, of cutting through the layers of hierarchy and avoiding the circumlocution of what had been considered the normal channels of communication. It was sad, Walker thought, that this had not been applied before.

Soon after Mountbatten's departure his nephew, Prince Philip, arrived, showing some of his uncle's characteristic intelligence, irritability and charm. What particularly irritated the Duke of Edinburgh was that, as the regulations laid down by London insisted that he never fly in single-engined aircraft over the jungle, he would only be able to visit troops stationed near airstrips where twin-engined light aircraft could land.

Prince Philip was the last of Walker's visitors in Borneo and he now set about making his farewell visits and handing over his command to General Lea.

Lea arrived during the Prince's visit and so, after a brief meeting with Walker, set off to tour the theatre for ten days. It was considered that visits made by the incoming Director of Operations on his own – meeting the brigade commanders, the staff and the battalions in the field – would be preferable to a tour guided by the officer he was to relieve. So it was that, after a comparatively short talk and a reading of Walker's hand-over notes, Lea was able to make up his own mind on the future conduct of the campaign.

At this time, most of the reinforcements had arrived and the formation of the new Mid-West Brigade headquarters was nearly complete. Lea saw that he was to inherit a miniature army, navy and air force complete in all its arms and appendages down from the infantry soldiers on the frontier to organisations for pay, welfare and public relations in the rear. All this was to be handed over to him but it was impressive, even awe-inspiring, to think that this force – involving perhaps twenty thousand men but without any 'surplus fat' – had been largely the creation of one man. Lea was clearly impressed by the scale of Walker's achievement.

Touring the frontier areas, Lea felt a sense of urgency for it seemed unlikely that the Indonesians would commit a strong force of their best troops to Borneo without a positive plan to use them. The intelligence reports available to him detailed the quantity of the Indonesian reinforcement, not their quality. Yet there was no reason to doubt that this was anything but high, although there might be shortcomings in the enemy's planning, administration and communications. And, at the back of his mind, there was always the internal threat and, although this danger was less obvious, it had led Walker to write at this time: 'The internal threat is sinister; it runs deep; it is underground. When it breaks, the external threat will be seen as child's play.'

Surveying this scene, Lea had mixed feelings. As a soldier, he was fascinated and stimulated. This was the kind of soldiering he understood, for he had commanded the SAS during three peak years of the Malayan Emergency and spent much time among the aboriginal tribes in the deep jungle. He felt at home in the jungle, he knew how to fight there and understood the importance of winning the hearts and minds of the people who lived there. And yet there seemed to be no direct and obvious means of achieving victory. The Indonesians had committed only a small part of their armed forces as yet and their reserves of manpower and weaponry and resources necessary for conventional war seemed vast. Walker had stopped them, and Lea could probably continue to do so, but how would it be possible to defeat them and end their confrontation of Malaysia? No end to the troubles was in sight.

When Prince Philip had returned to Singapore, Lea visited Walker for a final talk. Then, on 12th March, Walker formally handed over and Lea found himself on the island of Labuan responsible back through Singapore to Kuala Lumpur and London and forward, through Kuching, Sibu, Brunei and Tawau to the Indonesian frontier and beyond.

While Walker flew to Singapore for a farewell call on the Commander-in-Chief and awaited his ship to England, Lea was feeling the balance of the weapon that Walker had designed, forged, tempered and passed on. During the months that followed, he was to make few major changes. One of these was in the command structure. His personal relations with Major-General Peter Hunt were as cordial as Walker's had been but, had they not been so, their relative positions would have been impossible. As Director of Operations, Lea had total control over operations in the field, leaving little but administration to the Commander, Land Forces Borneo. But both men were glad when the Commander-in-Chief agreed that they could revert to the former system, with the Director of Operations doubling as commander of the troops he was directing.

Lea did not have to wait long for an indication of Indonesian intentions. On 27th April, a patrol base in the First Division was attacked. Like the old platoon bases, the company bases were themselves defended by only a third or less of the troops they supported, the majority being on patrol. So the company base at Plaman Mapu was manned by only a small rear party when a crack battalion of Indonesian regular parachute troops moved up to attack.

The 2nd Battalion, the Parachute Regiment, had arrived in Sarawak a month earlier to relieve the Argyll and Sutherland Highlanders and their initiation had been gentle. Indonesian activity had been noted but the fort at Plaman Mapu was held only by a company headquarters, a weak platoon of young soldiers and a mortar section, with a company sergeant-major in command. On 26th April, the Indonesians moved into position with great skill. Two assault forces, one of a hundred men, the other of fifty, armed with rocket-launchers and mortars, reached their start-lines undetected, while two more companies were deployed in reserve and support.

At five o'clock on the morning of the 27th, the assault began under a barrage of rockets and mortar-bombs. The Indonesians rushed the position, firing automatic weapons and climbing over the barbed-wire defences. Inside the base, fighting was at point-blank range, often face-to-face in weapon pits. In ten minutes the enemy had been driven back, only to reform and assault again. It was a battle of the intensity of fierce actions with the Japanese – or even of Rorke's Drift in the Zulu War – with each side matching the other's courage, the British sergeant-major at one point standing up and firing a machine-gun from the hip. This attack, too, was driven back and then a third. An hour and a half after the first assault, the Indonesians began to withdraw, taking some thirty casualties with them. The Parachute Regiment had lost two killed and eight wounded, including the cook who had fought alongside his comrades.

What was ominous about this battle was not only the strength and dash of the assault but that the Indonesians, instead of making their customary tip-and-run attack, had twice re-formed and charged again. It was now easy to imagine the possibilities of a battalion assault or, if the enemy had the organisational and logistical resources, the implications of an offensive on a brigade front.

Walker had met the external threat by cross-border operations – by attacking the Indonesians exactly as they tried to attack him; but doing so with infinitely greater efficiency – and now Lea continued his policy on a more intensive scale. Not only did his infantry battalions – British as well as Gurkha – dominate the frontier, they controlled a *cordon sanitaire* miles deep into Kalimantan. By mid-summer, virtually all contacts with the enemy were on his own side of the border and it was usual for a forward battalion to have its headquarters and one company in Malaysian Borneo and all other fighting troops inside Indonesian Borneo.

The utmost secrecy continued to surround these 'Claret' operations. They were never hinted at in news releases by Far East Command and, even at meetings of the National Defence Council in Kuala Lumpur, reference to them was kept to an essential minimum. However, when wounded soldiers were evacuated to hospital in Kuching, some announcement became necessary to pre-empt speculation, and a false cover-story was given to the news media. Stress had always been laid upon the importance of keeping trust with news correspondents and this was being broken, to the disquiet of some of those who had to take the responsibility. Nor was this the only drawback of such secrecy. Throughout South-East Asia – including Indonesia as well as Malaysia – it was assumed that the Indonesians were still on the offensive and that all the actions in Borneo were being fought inside East Malaysia, or Brunei.

So, if one of the Indonesian aims in military confrontation was aggressive propaganda, then this was being supplied for them by their opponents' caution and secrecy.

Thus, cross-border operations were conducted in a vacuum. No consideration need be given to the immediate political context, or to world opinion, so long as they remained secret. The Indonesians showed by their own silence that they had nothing to gain by admitting their own defeats, so long as these remained on an acceptable scale. Once they were hit too hard, and could provide evidence to prove that the British were violating their border, it might become worth their while to appeal to the United Nations in the hope that Asian opinion would support them and might succeed where their military commanders had failed. Thus the Director of Operations had to assess the likely Indonesian reaction to each attack and decide whether their reaction might be transformed from a military to a political one overnight.

So tight was security that a Gurkha could be awarded the Victoria Cross for gallantry inside Indonesian Borneo with a misleading citation. In November 1965, the 2/10th Gurkhas, on a 'Claret' operation opposite the First Division, stormed a strong Indonesian position on a razor-backed ridge killing at least twenty-four enemy for the loss of three themselves. In this assault, Lance-Corporal Rambahadur Limbu fought with such consistent *élan* and bravery as would qualify him in any war for the highest award. The risk that the location of his achievement would leak out was taken and he was sent to London to be decorated by the Queen. In the event, the full story of Limbu's VC was not reported until more than three years later.

But, while fierce company-scale actions continued – mostly inside Kalimantan – the political structure of Indonesia was crumbling. In October, 1965, a communist coup in Jakarta failed and this coincided with a successful rounding-up of CCO suspects in Borneo. As result, enemy activity immediately slacked, both externally and internally. Occasionally, there was a violent spasm of fighting. In March, 1966, a Gurkha battalion was involved in some of the fiercest fighting of the campaign when they made two raids into Kalimantan.

Lea had the measure of the most violent possible Indonesian action and his position had been strengthened, both politically and militarily, by the arrival of Australian and New Zealand troops. But, in the same month as these raids, Indonesia was convulsed with another coup, this time by the army with the aim of eradicating the growing power of the communists. Hundreds of thousands of suspected communists were killed and President Sukarno lost power in all but name to General Suharto. Now the Indonesians began negotiations with Malaysia and a peace conference was to meet in Bangkok. But it was not clear whether Suharto was in complete control and vigilance in Borneo could not be relaxed. In June, peace proposals were made by both sides. In July, Sukarno lost the last vestiges of his former influence. In August, the peace agreement between Indonesia and Malaysia was signed in Jakarta, the only Indonesian condition being that further elections be held in East Malaysia. On 11th August, the war officially ended – although there were some recent incursions to be tracked and dealt with – and at once the withdrawal of British forces began.

The Borneo campaign had lasted three years and nine months. At its height there had been about seventeen thousand Commonwealth soldiers, sailors and airmen in Borneo and another ten thousand in the South-East Asian theatre. Commonwealth losses had been one hundred and fourteen killed and one hundred and eighty-one wounded. Indonesian losses were officially put at about six hundred killed but, largely due to cross-border operations, the actual total was far higher.

Military opinion recognised the war to have been a classic example of the effective use of minimum force. New tactics had been developed, new lessons learned and old lessons re-learned to produce a coherent and cohesive policy that had succeeded brilliantly in startling contrast to the failure of the American policy – involving maximum force – in Vietnam. In the House of Commons, Denis Healey, the Secretary of State for Defence, said of the campaign: 'In the history books it will be recorded as one of the most efficient uses of military force in the history of the world.'

No Director of Operations could have asked for higher praise than that.

Shortly before his departure from Borneo, Walker had been made aware of the reception that awaited him in London. He had first heard of this at Kuching airport when he met Admiral Mountbatten. The two had just shaken hands when the Chief of the Defence Staff said that he was sorry about Walker's knighthood. Walker appeared uncomprehending and Mountbatten turned to Begg and asked whether he had been told. On hearing that he had not, Mountbatten said that he and the Commander-in-Chief, Far East, had both recommended Walker for a knighthood in recognition of his services in Borneo. Unfortunately this had not been approved by the Army Board.

What had happened was that this recommendation that Walker receive the accolade of a Knight Commander of the Bath had been rejected by Field-Marshal Sir Richard Hull, now Chief of the General Staff and soon to succeed Mountbatten as Chief of the Defence Staff. He did so because of what he regarded as Walker's disloyalty in the affair of the proposed reductions in the Brigade of Gurkhas more than two years before. A second proposal from Begg that Walker be appointed Commander of the Order of St Michael and St George was also rejected.

Finally, a suggestion from the Army Board that Walker be awarded a Bar to his Distinguished Service Order was received at Headquarters, Far East. It was pointed out that the DSO was a decoration for gallantry in the face of the enemy. Walker's first two had been won for this but, as Director of Operations, the DSO seemed inappropriate. The Army was adamant that the DSO was the highest award it would sanction and Begg finally agreed to put forward a recommendation, which was at once accorded.

The citation recorded that Walker had been 'constantly exposed to danger both from disaffected elements – especially during the Brunei rebellion – and from the hazards of flying vast distances in light aircraft. These flights frequently take place in very adverse weather conditions over inhospitable country and under the threat of enemy anti-aircraft fire.'

'But who in Borneo has not done that?' asked Walker when he read this. He was not so much angry, he told a few friends, as disgusted. Any award to a field commander was not only a recognition of that officer's services but also of the achievements of those under his command. On a personal level, a knighthood was an award that could be shared by a wife and this Beryl had richly earned. But he was particularly surprised that what had seemed to be a personal dispute with Hull over the Gurkha issue could have been carried on for two years and could be taken from Whitehall on to the battlefield.

Among the dozens of congratulatory letters sent to him were those from his own circle who knew the background and shared the reaction. Comments varied from such oblique comments on the DSO as 'a fitting but somewhat strange award' and 'a very small but typically British return for your work' to outright anger. One old comrade-in-arms wrote, 'I find it difficult to congratulate you on receiving an award which I consider to be meagre and shaming. I thought a year ago that you had been shamefully treated but hoped that the CB was only a necessary preliminary. Now I feel, as will so many others, humiliated and degraded in being connected, in however small a way, with those who have seen fit to recommend this miserable token of recognition.'

Outside the Brigade of Gurkhas and the Far East Command, the slight went generally unnoticed because the intensity of the campaign had been kept from the public. It was not until, two years later, as the Indonesian confrontation caved in after the collapse of Sukarno, that a discrepancy in rewards became apparent.

In the New Year Honours List for 1967, it was announced that Major-General George Lea, who had inherited Walker's campaign and carried its responsibilities so worthily, had been appointed a Knight Commander of the Bath.

7

Unarmed Combat

There was to be no conquering hero's welcome for Walker when he reached London and he had not expected one. It was not only that his victory – the victory that Lea would now complete – was being kept secret for diplomatic reasons. His was a new style of victory, the first of which had been won in Malaya. This did not involve a set-piece battle, the capture of territory and prisoners or a formal surrender by the enemy. It was a question of preventing the enemy from achieving his objectives to the point where these were abandoned. The Indonesians had not yet abandoned their hopes of taking East Malaysia and there were likely to be fierce battles ahead but it now seemed unlikely – even impossible – that they could achieve their ambition.

In many ways Walker would have preferred to remain in Borneo and, indeed, the Tunku had asked if he could stay. But he had been under pressure for a long time and understood the need for a rest to 'recharge his batteries' as he put it. Moreover, Begg had pointed out the importance to his future prospects of a senior staff appointment in NATO. Without an intimate understanding of the West's principal military alliance he could not hope for promotion to the highest positions in the Army. There seemed to be no obstacle to this now that his conflict with Hull seemed to have been brought to a close by the rejection of his knighthood.

But if Walker returned unnoticed by the British public, he expected at least full questioning from his superiors in the Army and in the Ministry of Defence. He had waged a successful campaign of this new sort.

He was the first general to experience the new unified command structure under war conditions and not even Admiral Begg had seen its workings at such close quarters at the operational end. He had developed new tactics using both minimum force and making maximum use of limited forces through air mobility. He had used almost the entire range of weapons and equipment – with the exception of heavy armour and missiles – that was available to the Army and had assessed its qualities. He had noted the capabilities of soldiers – British, Gurkha and Malay – under the most difficult conditions and his views might be the deciding factor in current theories about supporting the Far East theatre with quick reinforcement from Europe. Walker expected to face a lengthy and thorough 'de-briefing' in Whitehall before he could enjoy the six months' leave due to him.

This homecoming coincided with a period of upheaval in the Ministry of Defence.

After nearly a quarter of a century at the summit of defence policy-making, Mountbatten was at last to retire as Chief of the Defence Staff. In July, he was to be succeeded by Hull and this would be a difficult time to hand over. Mountbatten had seen the results of many of his decisions and although one of the more important of these, unified command overseas, had been seen to work well in South-East Asia it was by no means generally accepted by the majority of senior officers in the three Services. Far more serious a problem than this was the rekindling of the long-smouldering rivalry between the Royal Navy and the RAF over the command and control of air power over the sea.

The division of responsibilities in the air between the three Services had been a cause of perennial squabbling since the First World War and had led to the adoption of the rigid dogma of the indivisibility of air power by the RAF, which Walker had had to contend with in Borneo. The crux of the question was the Fleet Air Arm. The RAF suspected that the Navy wanted to control its own maritime aircraft of Coastal Command and the Navy was convinced that the RAF was hoping for the share of the defence budget now devoted to its aircraft

carriers. Although the Conservative Government had, in 1957, authorised at least one new carrier, this question-was being re-examined by
their Labour successors and lobbying, intrigue and argument between
the two Services had become intense.

Walker did not believe in the indivisibility of air power and recognised the need for carriers. It was thought that Mountbatten held the
same view but now the decision on their future would be taken under
the direction of Hull who had had no experience of maritime war.
Thus, Walker did not expect the Navy and RAF to show the same
interest in the lessons of the Borneo campaign as the Army – free of
such preoccupations – would doubtless demonstrate.

Soon after his return to London, Walker was summoned by Mountbatten. After congratulating him on his achievement and adding that a
NATO appointment was now essential to his future career prospects,
Mountbatten asked about unified command. He was still facing problems in the new structure in Singapore and wanted Walker's views.
Among these were that the three separate Service headquarters,
which still survived on a massive scale in Far East Command, were
now redundant and that a small, streamlined, unified headquarters
would be more efficient and less expensive.

Next Walker saw Healey, who repeated Mountbatten's words
about the importance of NATO experience. Clearly their last meeting,
in Borneo, two years before, had made a deep impression. He told
Walker that two points he had made then had stuck in his mind. One
was that a battalion with helicopters was worth a brigade without. The
other, that armies should be organised from front to rear and not from
rear to front: that priority should be given to giving the best commanders and equipment to the fighting troops and not to the headquarters
staff. He, too, wanted to know about the success of unified command
and suggested that Walker should get down to detail with his Deputy,
Fred Mulley.

Walker's high regard for these two Labour Ministers – so much higher
than that for their Conservative predecessors – was now confirmed.

Mulley he found unshaven, tired and more untidy than usual after an all-night sitting in the House of Commons, yet his interest and enthusiasm was much alive. After more questions about unified command, Mulley ranged over a wide variety of defence problems. Walker's replies were always astute, often unexpected. For example, a question about the relative merits of young officers with different training. The best in Borneo, said Walker, were firstly those produced by the Royal Marines, secondly by the Army's short-service commission training and thirdly by the Royal Military Academy, Sandhurst, in that order.

Mulley listened intently for more than an hour, then asked Walker if he would put his ideas on paper. This would not be easy, Walker realised, because he expected an immediate series of de-briefing interviews with the Army and the Army Board would expect any such paper to be presented first to them. On the other hand, it seemed probable in the light of experience that only those opinions that supported their own would be forwarded to the Ministers and to the Chief of the Defence Staff. Walker therefore agreed to preparing a private report to be sent directly to Mulley and Mountbatten, by-passing the Army Board.

The Walkers had taken a flat in Kensington to be near Venetia, who was at a secretarial school. The report was written there and typed by Venetia. When it was complete Walker handed the two copies to an officer on Mountbatten's staff in the entrance hall of the Ministry of Defence to avoid an embarrassing meeting with Army aquaintances in the corridors of power upstairs.

The next expected summons would surely be from the Army. But none came. No member of the Army Board wanted to hear about the Borneo campaign from its Director of Operations, any more than did the Foreign Office or the intelligence organisations. Nor were there any social invitations; clearly hospitabilty in London was not so generous as his had been in the East.

When a summons did arrive it was from Field-Marshal Sir Gerald Templer, who was conducting an enquiry into the rationalisation of air power and asked for evidence from Walker on the command and control of aircraft – including helicopters – in Borneo. Two days after this request was received Walker heard from the Army for the first time. The CGS required that, before giving his opinions to Templer, Walker be instructed in their views on the subject. Walker replied that Templer surely did not want the official Army view, which he doubtless knew already, but had asked for the personal opinions of the former inter-Service Director of Borneo Operations in the light of his recent operational experience.

The Army's reaction was sharp. Walker was told that on the instructions of General Sir James Cassels, who had just relieved Hull as Chief of the General Staff, he was to read a directive from the Army Board on air policy. It was an order and Walker complied. He then assembled his own opinions – stressing the need for decentralisation and flexibility in air operations supporting ground forces – and appeared before Templer's committee of enquiry to give them.

This was the first and last contact with the Army Staff following his return from the Far East. He had powerful friends but none of them seemed to be soldiers. They were Mountbatten (soon to be out of office), Healey, Mulley, Begg, who was soon to be succeeded in the Far East by Air Chief Marshal Sir John Grandy, and a mixed collection of senior officers in the Royal Navy and RAF, and civil servants, mostly retired or about to retire.

Yet he had one opportunity to make his voice heard. Brigadier Patterson was now a student at the Imperial Defence College and, at his suggestion, Walker was invited to lecture there as the first general to implement unified command in war. This was a chance to repeat his views of the importance of organising the Services from front to rear and he told the students – all of them being prepared for senior appointments – that too many good officers were either on the staff or being trained for it, poring over books when they should be working with their soldiers.

He was able to speak freely again when invited to address the Joint Services Staff College and the Army Staff College by their Commandants.

But if Walker was neglected in London, he was remembered by old friends in Malaysia. In June, he received a telegram from the Prime Minister's office in Kuala Lumpur asking if he would accept the award of Honorary Penglima Mangku Negara, the Malaysian equivalent to knighthood. He accepted and was told that the investiture would be by the Tunku himself during his visit to London in September. So while Walker was not 'Sir Walter' he would now be entitled to call himself Major-General Tan Sri Walter Walker, an anomaly in the balance of rewards and services that reminded more than one of his friends of the awards to Nelson after the Battle of the Nile: a barony from his own country and the Dukedom of Brontë from the King of Naples.

The Walkers spent their six months' leave in London and, early in September, left for France and the intricacies of the North Atlantic Treaty Organisation. That this was to be a totally different life from that of the past years, both professionally and personally, was clear. Instead of stalking jungle infiltrators and thwarting them with the minimum use of force, Walker's military task would now be to prepare to meet attack on the West by the most powerful army in the world using massed air power, armour and probably nuclear weapons. The task was to help hold off or, failing that, win the Third World War. The enemy would not be an egomaniac dictator but the sombre, calculating leadership of the Soviet Union.

In this new and vastly larger theatre, Walker would still be planning operations but he would be one of many with such responsibilities. He would also be working in a new context: among officers of a dozen nations of whom a small minority would be those among whom he had served. Indeed, he must no longer regard himself as an officer in the British Army; now he belonged to the multi-national legions of NATO.

Since it had been formed to meet the threat of further Russian expansion in 1951, the military structure of NATO had developed into a complex of interlocking commands. Of these the most important, militarily and politically, was the Allied Command Europe, facing the

Russians in a wide arc from eastern Turkey to northern Norway with the Channel and Atlantic Commands guarding its lines of communication with Britain and North America. In Western Europe, the sector demanding the heaviest defences was the traditional invasion route from the east: the plains of Germany between the North Sea and the Alps. Here were drawn up the Allied Forces Central Europe: British, American, French and German armies and smaller military contributions from the Netherlands, Belgium and Canada with tactical air forces and missiles in support, known as AFCENT.

The threat from the east seemed at least as real as it had a quarter of a century before, although – due to the deterrent of NATO and its massive support in American strategic nuclear weaponry – the Russians had not extended their occupation of Eastern Europe one inch to the west.

The fifth American officer to be appointed Supreme Commander, General Lyman L. Lemnitzer, had his headquarters – Supreme Headquarters, Allied Powers Europe, or SHAPE – at Rocquencourt outside Paris. Allied Forces Central Europe, commanded by the French General Crépin, had its headquarters at Fontainebleau. The main headquarters and that of the land forces were in the Chateau de Fontainebleau itself and air headquarters in a barracks complex nearby.

The Commander-in-Chief Land Forces was the German General Graf Johann von Kielmansegg, who had already served at SHAPE as representative of West Germany and the Commander-in-Chief Air Forces was the British Air Chief Marshal Sir Edmund Huddleston.

At both, large staffs – manned and equipped on the lavish American scale – had developed, drafted and revised plans to meet all conceivable contingencies. But while the plans themselves, the products of some of the best military minds in the Alliance, were of the high quality to be expected, the succession of officers in the various headquarters were of variable worth. Indeed it was impossible to maintain consistent standards on the staffs, if only because it was virtually impossible to get rid of an officer, once appointed, until his term of duty was complete.

Many officers were, of course, appointed to NATO on grounds of professional ability. Others, however, were recommended for duty at Rocquencourt or Fontainebleau as a reward, or as a posting convenient to promotion schedules, or even to keep them away from their own countries. For an officer, once accepted by NATO, to be sent home was fraught with such diplomatic risks that all but the most damaging misfits had to be left where they were, and work of which they were incapable given to others. Walker, remembering the small, highly-professional, hard-working staff of uniformly high quality he had had in Borneo, realised that this would be difficult, if not impossible, to achieve in NATO.

It was not in Walker's nature to regard duty with NATO as anything but work demanding the maximum concentration. If the aim was to deter Russian aggression, then they must realise that NATO forces could actually deter and were not just stage scenery masking American nuclear power. Walker's own contribution to this could only begin after he had mastered the current intelligence assessments and operational plans. This could not be done only by briefings at headquarters and by reading, he was convinced. It involved touring the areas where the NATO armies might have to fight, talking to their commanders and getting the feel of the terrain. At one of these early meetings, General Kielmansegg agreed that this would be useful.

Walker was not accustomed to being part of those rambling headquarters he so despised, or, for that matter, to being unable to get his own way in what he considered a matter of operational importance. He was therefore shocked and angered when Kielmansegg's Chief of Staff, the Belgian Lieutenant-General Jean Ducq, told him there was too much work pending for any tour to be allowed. There was a tense confrontation, but Walker was out-ranked and had no alternative to accepting the order, albeit with poor grace. He returned to his office, remarking that he would be nothing more than 'a bloody babu'. It was four months before his relations with Ducq relaxed, and then, when the two came to know and understand one another, they became friends.

One of Walker's first tasks at Fontainebleau was to help prepare for an exercise called FALLEX 67 designed to test the functioning of the command structure, the machinery for decision-taking and the speed of reaction. He had already been critical of the over-manning of the staff and the verbosity of its American procedures, but now he was appalled.

The crux of NATO strategic problems was the taking of the decision to use nuclear weapons. This must obviously be a last resort, the aim being to hold any Russian aggression with conventional forces for long enough to allow politicians in both East and West to appreciate the situation and its dangers and the consequences of escalation. There could be no question of holding a determined Russian offensive for more than a few days without resorting to tactical nuclear weapons on the battlefield and in the support areas, but these days would decide whether peace could be restored, or whether escalation to mutual bombardment with intercontinental strategic nuclear missiles was to begin.

The most vital ingredient of this terrible sequence of decisions was clarity and speed. The politicians must know as quickly as possible of developments on the battlefield and the commanders must know as quickly as possible whether or not a nuclear response was to be made. From the purely military point of view it was vital that such orders should be received almost instantly, if only because the situation would change so fast, and the pattern of targets would change. In this process, Walker was in the key position. Reports from field commanders would be passed from him to AFCENT, and to SHAPE, and he would also deliver the consequent orders to the forces on the ground.

So it was with amazement and horror that he found that the most urgent messages, from both the Army Groups in the field and from

SHAPE, took an average of four hours to reach his desk under simulated war conditions. Reporting this to Kielmansegg, Walker said that he had had better communications on the North-West Frontier of India, using signal flags and heliograph reflectors by day and morse lamps by night, than now when the most sophisticated electronic systems were available on a lavish scale.

Kielmansegg at once gave him a free hand to reorganise communications. As a start, he designed a mobile headquarters in caravans. His own, which was out of bounds to all except a small staff, had map displays and a small cell of transparent Perspex, where he himself could receive and make signals, see the maps and remain in radio contact with the commanders in the field and with the main headquarters. By cutting out all inessentials, he reduced the four-hour delay to twelve minutes and was able to contact Kielmansegg and Huddleston within seconds.

Walker based his new staff system on the methods he had learned during the retreat from Burma when, as one of Slim's staff officers, he had had to act quickly. Now he cut out all superfluities to reduce to the minimum the number of stages through which messages had to pass. In remembering the Spartan simplicities of staff work in Burma, he realised that there were similarities between jungle warfare and what could be expected after an initial nuclear exchange. In both, the ponderous machinery of logistics and infrastructure could break up, leaving each individual on his own or with rare, brief contacts with a central command. Therefore, understanding between commanders in the field and at headquarters must be close and initiative intelligently used. So, from the mountains of Waziristan and the jungles of the Irrawaddy valley, military practicality reached Western Europe.

In 1966, NATO was to be tested, but not by the Russians. France, the geographic keystone of the defences, had been becoming an increasingly restive member of the Atlantic Alliance and President de Gaulle was challenging American domination of Western Europe. In February of that year, the President announced that all military forces in France were to come under French command. French forces were therefore to be withdrawn from participation with NATO by 1st July and all NATO forces – including headquarters and infrastructure, would have to be withdrawn from France by 1st April, 1967.

The strategic consequences of the expulsion – involving the loss of defence in depth and the concentration of all NATO land forces in the forward areas, in West Germany and the Low Countries – were serious enough, but they had to be accepted. The immediate problem was the uprooting of the headquarters, the communications systems, the airfields, stockpiles, fuel dumps and pipelines from France, where they had been slowly assembled in vast quantities over twenty-five years, and re-establishing them farther east. Moreover, this had to be done not only quickly but without leaving a vulnerable space of time when the defences would not be fully functional, so offering a temptation to the Russians to undertake some aggressive adventure.

At this time, the problems of the exodus from France did not directly concern Walker and he attended a senior officers' course in nuclear warfare at Oberammergau in Bavaria. There he made friends with the Italian Lieutenant-General Umberto de Martino, who commanded NATO Land Forces Southern Europe. Walker had been talking about Gurkha infantry and de Martino had remarked that his Italian alpine troops were of the same quality. The two became friends and Walker was invited to watch an exercise in the Julian Alps that summer.

Leave was due, so Walker took his wife to Italy and they stayed at Verona, where he could visit de Martino's headquarters and fly with him to the exercise area. He was looking forward to mixing with soldiers again – particularly if they were anything like Gurkhas – and in shaking off the confines of a headquarters in mountain air. Just as Walker was leaving for the airfield at Verona, a packet of mail, forwarded from Fontainebleau, arrived, but there was no time to open any of it before boarding the helicopter. So it was not until he was strapped in the aircraft, and flying towards the ranges of mountains that began to lift out of the plains of North Italy, that Walker noticed one envelope marked 'Private and Confidential' and bearing the insignia of the Military Secretary's office at the Ministry of Defence.

Lieutenant-General Sir Richard Goodwin, the Military Secretary, began his letter with 'My dear Walter' and signed it 'Dick', but the words in between Walker read with amazement. Goodwin was telling him, obviously with reluctance and embarrassment, that he was to be retired from the Army. There was to be no promotion and, as there were so few vacant appointments for major-generals, it was a question of 'either "up" or "out"'. Goodwin concluded that, although No. 1 Selection Board had not yet made its decision final, he had to warn Walker that it was most probable that he would have to retire between March and September, 1967.

The letter had a ring of finality about it because Walker recognised that he had few, if any, allies among the generals in Whitehall. The last member of No. 1 Selection Board with whom he had been in close contact, Hewetson, the Adjutant-General, had, just before he had left the Far East, referred to Walker's 'lack of loyalty to Khaki'. Although he had been on friendly terms with the new CGS, Cassels, when he had been Director of Operations during the Malayan Emergency, he could not expect support from his predecessor, Hull, who had now succeeded Mountbatten as Chief of the Defence Staff.

It was clear that what was now called 'The Establishment' of the Army was determined to get rid of him despite the assurances from Mountbatten, Healey and Mulley and the loyal support of Begg. There was unlikely to be a response to an appeal to any of his seniors in the Army.

Walker had to stifle these thoughts. General de Martino was speaking to him on the inter-com. system, describing the country the helicopter was crossing. He put the letter in his pocket and its contents out of his mind. For the rest of the day he showed keen interest in the impressive display and parade by the alpine division. It was not until his return to Verona that night, when he showed Goodwin's letter to Beryl, that he began to make plans to defend himself.

The chance of a successful appeal against the confirmation of the Board's decision seemed at best minimal. If they had decided that Walker could not be a candidate for a major-general's appointment, it was hardly likely that Cassels would consider him for promotion to lieutenant-general. Outside the Army hierarchy, there seemed little hope of finding an influential ally. In 1965, Fred Mulley had become Aviation Minister and been succeeded as Army Minister by the young Gerry Reynolds, whom Walker did not know. The only man who could overrule the Army, and might be willing to do so, was Denis Healey, but to make a direct approach to the Secretary of State for Defence over the heads of all his superiors would be so glaring a breach of protocol that that act alone would be enough to destroy any hope of a continuing career.

On his return to Fontainebleau, Walker noticed a newspaper report that Fred Mulley was about to visit Paris for an aviation conference. At once he passed a message to the Ministry of Aviation asking whether, as senior British Army officer in the Paris area, he might make a courtesy call on the Minister at his hotel. Mulley readily agreed and suggested he come round for coffee after breakfast on 27th July.

It was a pleasant, informal meeting. Mulley talked about his new job and, after asking Walker about his, enquired whether he had yet heard of his next appointment. Walker then told him about the Military Secretary's letter. Mulley was horrified. Could he help? He would see Reynolds and Healey immediately he returned to London and ask them to intervene. Walker replied that he would naturally be grateful for any help that Mulley could offer.

Two days later, Walker replied to the Military Secretary's letter. It was a formal representation against the decision of No. 1 Selection Board. He quoted the assurances from Healey, Mulley and Mountbatten that NATO experience was necessary for his future career and he quoted comments from congratulatory letters written to him by his superiors in the Far East. He also quoted a report in the London

Evening Standard, which he assumed to have been inspired by the Ministry of Defence. This, after describing him as 'the most brilliant and experienced jungle war expert in any army', said that 'his success in Borneo has marked him for the highest commands and, for these, experience in Europe is essential'.

His letter concluded by asking why he had been singled out for retirement and he maintained that his treatment was 'unfair in the extreme.' He asked Goodwin to represent his case to No. 1 Selection Board.

Despite the Military Secretary's diplomatic pleading that there was an acute shortage of employment for senior officers, Walker was convinced that he had been singled out for premature retirement. Of nine contemporaries, who had recently been promoted to lieutenant-general, only one – Michael Carver who had succeeded Jolly as Army Commander in the Far East – had commanded a Regular division as he had.

At the beginning of August, Walker received an acknowledgement to his letter from the Deputy Military Secretary and the assurance that his case would be considered by the Board when it met on 5th October. There were exactly two months in which to save his career.

If Walker was to fight back, then more allies had to be found and it was important that as many friends as possible knew of his danger. Few would be able to help directly but they might create a climate of opinion useful to him. Failing all else, influential friends might help him find suitable employment outside the Army. One of the first letters he wrote was to Varyl Begg but he began this by recognising that his old friend was heavily involved in the affairs of his own Service. Begg was now First Sea Lord. This had long been expected but, in 1966, the quarrel between the Royal Navy and the RAF over maritime aviation had come to a head. After months of acrimonious debate, Denis Healey had stepped in to support the RAF and had cancelled the proposed aircraft carrier programme. Both the Navy Minister, Christopher Mayhew, and the First Sea Lord, Admiral Sir David Luce, resigned in protest. For the second time Begg succeeded Luce but, on this occasion, in the midst of a political crisis. His main preoccupation was now to restore the Navy's shaken morale.

Although Begg would not be in a position to intervene directly, he was a popular and influential man with even more influential friends, one of these being Mountbatten. Walker also wrote to old friends who had retired from the Far East, making his first, tentative enquiries about a new civilian career.

Among sympathetic and sometimes helpful letters that began to arrive was a short, hand-written note from Mulley to say that he had spoken to both Healey and Reynolds about his future and that they had promised to 'look into it'. A fuse had been lit and there was now no more Walker could do but wait. One distraction came in September when he visited London to receive his accolade from the Tunku at the Malaysian High Commission in Belgravia. The Malaysians, at least, were grateful for what he had achieved.

This was a particularly welcome gesture because Walker's appointment at AFCENT was now to end almost a year early. The expulsion from France was to be preceded by a major reorganisation and streamlining, necessitating a redistribution of staff appointments to the various nationalities. An American was to become the new Deputy Chief of Staff (Operations and Intelligence), leaving Walker redundant.

Essentially, the reorganisation was to merge the land and air headquarters into one unified command – the naval headquarters had already gone – on the lines recommended by Mountbatten. Kielmansegg would become overall Commander-in-Chief and his new, smaller combined headquarters would be formed by 1st December, exactly four months before it would have to start operating outside France.

With reluctance, Walker began to make his personal contingency plans. He had put out feelers in the Far East on the chance that there might be work for him in one of the British High Commissions, or as a defence adviser, and now turned his attention to the possibilities in Britain.

Among the civilian jobs he enquired about was one as general manager of the Skelmersdale Development Corporation in a 'new town' for Mersey-side and as a senior administrator for the National Coal Board. It was suggested that he put his name down for employment with the Home Office.

No further word had come from Mulley but their meeting in Paris had, in fact, already achieved its aim. Healey had been surprised and shocked that one of the senior officers he regarded as the most promising was to be retired. In September, he informed the Army Board that the future of General Walker should be reconsidered. On 5th October, No. 1 Selection Board met and Goodwin gave them Walker's letter of protest. About a week later Walker was informed that the Board would consider him for promotion with others for appointments as they occurred. It was stressed that competition would be intense and there were no promises. Yet compulsory retirement had been staved off, even if only temporarily.

Meanwhile Kielmansegg found him more work. It had been decided that when SHAPE moved to its new site at Casteau, near Mons in Belgium, his own headquarters would go to the Netherlands. The only possible accomodation that had been found was in the offices of a coal mine at Brunssum in the northern province of Limburg. An advance party had been sent there but the task of moving a functioning headquarters with all its equipment and communications, the staff and their families, the transport and the war headquarters, in three months, might well be beyond them.

When Kielmansegg visited Brunssum he was convinced that the move could not be completed on time or, at least, that, even if the personnel and equipment could be moved, an operational headquarters would not be functioning there by 1st April. So he appointed Walker as his Director of Movement Coordination, to be in complete control of the move. Although this would involve a rearrangement of the staff, possibly involving hurt feelings, he was to be given a free hand.

Kielmansegg could not have made a more suitable choice. Since he had been a young staff captain on the North-West Frontier of India, one of Walker's specialities had been the rapid moving of headquarters and camps. What he had done in the hills of Waziristan was now to be repeated on a massive scale. Now he was allowed to show the same ruthless efficiency and drive as he had in the Far East and was soon to earn the nickname, 'The Dictator'.

Walker treated the move to Brunssum as an operation, knowing that his decisions would be supported by Kielmansegg. When he arrived at the new site, miners were still using the showers and changing rooms and coal trains were still running through what was to become the headquarters area. The office blocks would have to be gutted and refitted and new buildings would have to go up. Accomodation would have to be found for a staff of some six hundred and their families. Meanwhile, the headquarters at Fontainebleau would have to be evacuated and handed back to the French in good order, as would the mass of hired accomodation in the area. This, too, was Walker's responsibility.

As the buildings were converted and new ones built, Walker moved his own office almost daily, working to a strict timetable. Within three months and well before the end of March, the headquarters of AFCENT had moved into offices that were functional if not always comfortable and, before the end of the month, Kielmansegg arrived for the inauguration ceremony. Not for a moment had the headquarters ceased to operate fully in readiness for any emergency. The Commander-in-Chief told Walker that he had 'achieved the impossible'.

After these congratulations, Kielmansegg added that he would make it his business to see that Walker's achievement was recognised in the highest places, including those in London. Meanwhile his next task for AFCENT would be to prepare for a major exercise that was held every two years. This one, Exercise Lion 67, involving the Northern and Central Army Groups, representing NATO and Warsaw Pact armies, would be one of the largest exercises of its kind.

Yet, for six months, there had been no further word from the Ministry of Defence about his future. It was always possible that No. 1 Selection Board had obeyed Healey's wish to reconsider his future but, having done so, decided that there was no work for him after all. Or he might be found a job away from the main stream of his profession.

This possibility seemed to be the case when General Cassels, the CGS, telephoned from London. Had Walker read the Mountbatten Report on Prisons, which had been urgently prepared following the dramatic escape of the Russian spy George Blake from Wormwood Scrubs? Walker's name had been put forward – probably by Mountbatten himself – as a possible candidate for the new post of Director of Prisons and the CGS was in favour of him taking it. Candidates were to be interviewed by the Home Secretary, Mr Roy Jenkins, who would like to see him in two days' time.

Walker flew to London and, at the Home Office, met Jenkins. The utmost courtesy was shown by both but it was clear from the start that they did not warm to each other and that it was unlikely that Walker would be offered the job or, if he was, that he would accept it.

Jenkins wanted to know whether Walker had had any experience of prisons. He had guarded Japanese war criminals in Malaya, he replied, when they had hanged six before breakfast most mornings and he had guarded the Indian National Army prisoners in Bangkok. Then there had been the communist terrorists detained during the Malayan Emergency when none under his control had escaped. How had this been achieved? By fierce guard-dogs roaming free between belts of wire and watch-towers manned by local forces – there had been no waste of soldiers there. Jenkins wanted to know how Walker would prevent escapes from prisons in the middle of towns. By closing the surrounding streets. If Walker was selected as Director, what would be his first requirement? A helicopter, so that each prison could be visited regularly, because he would behave as a commanding officer and the staff of the Prison Service would be his troops. In fact, he would treat the whole task as a military operation.

It was no surprise to Walker when told that he had not been chosen. He had already told friends that he considered Jenkins not only uninspiring but cold.

Returning to Brunssum, Walker was preparing to hand over the preparations for the exercise to others and go on leave, when he received another letter from the Military Secretary. This told him that he had been appointed the next General Officer Commanding-in-Chief, Northern Command, in the United Kingdom with the rank of lieutenant-general. The Army's decision had been reversed – presumably by pressure from Healey, so now there would be at least one more senior appointment. Gratified to have challenged 'The Establishment' and won, Walker was both elated and depressed. The experience had shown that hitherto Ministers had either been by-passed in the allocation of appointments or had not ventured to interfere. He knew many fine officers who had been prematurely retired and had accepted their fate without a protest. Perhaps if they, too, had fought back they would now be in the highest ranks, where he thought they had belonged, in the place of those who had tried and failed to get rid of him.

Congratulatory letters poured in and several of those from his closest friends made the point that, while the new appointment might not make the best use of his talents, it at least 'kept him in the game' and ensured him of promotion and, almost certainly, the knighthood that would, as Mountbatten had put it, 'come up with the rations'. Walker was to relieve Lieutenant-General Sir Geoffrey Musson – nicknamed 'Daddy' because of his venerable appearance – who was to become Adjutant-General, in October, 1967.

To most conventional senior officers, Northern Command was one of the most sought-after prizes. Although accorded the title of 'Army Commander', the GOC-in-C had no army to command and was principally the landlord and administrator of those troops who happened to be stationed in the northern counties of England. With Command Headquarters in York and three District Headquarters under major-generals, the GOC-in-C could do as much, or as little, work – within reason – as he cared.

For those with social and sporting inclination, Northern Command was a paradise. These counties were the strongholds of the old landed aristocracy, untainted by the soft-living of the south, where it was the natural and honourable course for heirs to devote their lives to their estates and to field sports. The possibilities for hunting, shooting and fishing were limitless and the Army Commander would always find himself a lionised guest at the grandest dinner tables. Thus the GOC-in-C was expected to live in comparable style and the Walkers found themselves moving into Claxton Hall, near York, which was reputed to be the most elegant Army Commander's house in the country.

A few weeks after the Walkers had settled into these comforts, another official letter arrived from the Military Secretary. This one was to announce that Walker's knighthood had indeed 'come up with the rations' or, as General Goodwin put it, that the Queen proposed to honour him in the New Year Honours List with the award of Knight Commander of the Order of the Bath. Among the letters and telegrams that arrived in stacks, after the public announcement, were several from old Far East friends, adding words such as 'Three years too late!'

Letters of congratulation for automatic honours usually had a hollow ring to the heartiness but, on this occasion, there was the knowledge that most of the writers felt that this knighthood had less to do with the reaching of a certain seniority than with the conduct of the campaign in Borneo. A hand-written letter from Mountbatten, sent from Sandringham, implied as much.

Walker himself regarded his knighthood as a recognition to be shared. Beryl would, of course, share the title but, personal gratification apart, this was, he felt, also a gesture of gratitude to the Brigade of Gurkhas. It never had been – and never would be – acknowledged that Walker's stand against the plan to dispense with the Gurkhas had been vindicated by their decisive part in the defence of Malaysia. This knighthood would be the nearest to a public vindication that could be expected.

Accustomed to working under self-imposed pressure, on the assumption that the only valid prediction was that the unexpected was to be expected, Walker's first sight of Northern Command was not one to summon up the blood. He was particularly irritated at being described as an Army Commander and not only having no army, but not being part of the Army's chain of command. Operational troops had, in fact, recently arrived at Catterick, the vast barracks area in the North Riding, when 6 Infantry Brigade was withdrawn from Germany for economic reasons and stationed there. But Walker was only their landlord.

Of far greater concern than his own disappointment at this lack of responsibilities was the low state of morale he found among the soldiers in Northern Command. This was not exclusive to this particular area, nor to the Army but, he heard, was widespread throughout the three Services. It was the result of the unpredictable and erratic defence policies of the Labour Government over the past three years. Like many other professionals, Walker had welcomed the appointment of Denis Healey as Defence Minister, had come to like him and his junior Ministers personally, and had found their approach fresh and stimulating after the succession of Conservative Ministers with whom he had had to deal. But it was becoming increasingly apparent that they were far from being their own masters and that, since, in the Labour Party as a whole, the subject of defence was anathema, the Government's reflex action when economies were required was to turn to the Services, axe in hand.

The basic uncertainty was over the future role of the Services. After hesitation and changes of course, the Government had decided to withdraw from nearly all responsibilities 'East of Suez', by which was meant all but the European and North Atlantic theatres. Britain was hoping to join the European Economic Community and, as such, would become just another European state, past burdens and glories finally cast aside.

Northern Command could provide the best possible cross-section of opinion. Since the arrival of the brigade from Germany there were now 'teeth arms' as well as supporting units and training establishments and there were, of course, none of the distorting factors – good and bad – of service overseas. Talking with young officers and NCOs during his first tours of the Command, Walker found an almost universal lack of faith, not only in politicians but with the highest echelons of the Army itself. Repeatedly he heard cynical comments about politicians' promises and about lack of communication with their own leadership to which they were looking for reassurance and guidance.

Walker hesitated before approaching General Cassels, who was reaching the end of his term as CGS. Walker had admired him as a field commander but now it was said that he was tired and his relations with Healey were thought to be far from happy. He would welcome retirement, it was being said.

Attending a meeting of the Army Board with other Army Commanders, Walker spoke of the failure of the recruiting campaign and the Government's reluctance to support their Army. As a start, could not Cassels broadcast to the Army – or even to the nation – just as Slim had, when CIGS, to define their role in the future and to tell them that they remained as important to national life as ever they were? Cassels was sympathetic to Walker's case but the gist of his reply was: what can I say?

Cassels was to be succeeded by General Sir Geoffrey Baker – known as 'George' Baker to the Army – and Walker hoped that, coming fresh to the scene, he would now give the leadership that was so desperately needed. Northern Command would be an ideal setting for a display of such dynamic leadership not only because of the number of troops stationed there but because a high proportion of recruits came from the North and encouragement could be disseminated through families, both to soldiers serving elsewhere and to potential recruits.

Increasingly, Walker saw that his principal constructive task in Northern Command was to be recruiting. When Denis Healey arrived at York on his way to visit Catterick he asked whether Walker had any points he wanted to make directly to him. 'Yes,' replied Walker, 'recruiting.'

Healey then listened intently to what Walker called his 'Ten Commandments' for reviving the health of the Army. The inspiration for these had been talks Walker had had with his sons' friends, young officers in their twenties, whom he invited to be brutally frank in their comments. He had also made a point of questioning NCOs who were, of course, in the most intimate contact with the soldiers.

The first 'Commandment' resulting from such conversations was that the minimum six-year engagement for recruits should be cut to one of three. The reasoning was that, however eager a young man might be to join, his girl friend would see six years as a very long time to be committed to an adventurous but unpredictable way of life. Three years, on the other hand, would not seem an interminable commitment and, Walker had been repeatedly assured, three-quarters of soldiers engaging for three years would sign on to complete six. This scheme had been considered by Whitehall but had been turned down as impracticable from an administrative and training point of view. Now, Walker urged urged Healey to reconsider it.

A major factor in the need for a three-year engagement was the increasing tendency for soldiers to marry young and Walker's second 'Commandment' was that marriage allowances should be payable before the age of twenty-one. Another point was that, before leaving the Army, a soldier should be given a full course of training for resettlement in civilian life as an integral part of his Service career.

Walker had several points to make about methods of recruiting. Its organisation should be decentralised from London to the Commands and Districts. The recruiting officers should not be older men near retirement but able young officers who had seen active service – and why should they not be paid by results like insurance agents? Public relations should become a more important part of recruiting and the public must be shown that the soldier was now a highly-skilled professional.

Healey noted Walker's suggestions and promised that he would examine them in detail. A few days later, General Musson, Walker's predecessor in York who was now Adjutant-General, was seen moving at a jog-trot down a corridor of the Ministry of Defence. Asked what was the emergency, he is said to have panted, 'Walker's bloody Ten Commandments.'

In March, 1968, Baker succeeded Cassels as CGS and, on arrival at his office, found a letter from Walker waiting on his desk. This began as formal congratulations and a welcome, then set out his ideas for reform and, later, this was followed by a full report on his plans. Baker appeared to be as interested as Healey had been and, within a year, several of Walker's suggestions – including the three-year engagement and earlier marriage allowances – had been adopted.

Meanwhile Walker himself staged a spectacular public relations exercise, making use of 6 Brigade at Catterick. This was a mock battle, involving the new heavy Chieftain tanks and helicopters and other new weapons and equipment, lasting an hour and a half and impressing the crowds as the largest and most realistic demonstration of this kind seen in Britain since the war. For some of the young soldiers taking part it was their first taste of realistic, simulated, action.

Through his preoccupation with recruiting and his public relations activities, Walker had what was to prove a fateful meeting. A speech he had made on recruiting, in which he warned that, unless it improved, it might be necessary to reintroduce conscription, was reported in the *Yorkshire Post*. It was read by Malcolm Campbell of Tyne-Tees Television, at Newcastle-upon-Tyne, who thought that this outspoken general might be a lively subject for a television interview. Tyne-Tees had a reputation as the best of the commercial companies in the documentary field and, as Head of News and Current Affairs, Campbell had an original and stimulating approach to his subjects. In this case, he decided that Walker should be interviewed, not by a reporter with some knowledge of the Army and respect for senior officers' but by a brash but intelligent young reporter called John Hobbs, who could be expected to take a sceptical, even abrasive, line.

Hobbs telephoned the public relations office at Northern Command Headquarters and asked whether Walker could be interviewed about recruiting on television, together with a private soldier. He was told that Walker could not possibly appear with a private and that he would be told in due course whether or not the GOC-in-C would consent to appear at all. Three weeks passed, without word from the public relations officer, and Hobbs decided that 'the story is irrelevant, this is a challenge'.

John Hobbs finally went to York and asked to see Walker. He was passed from one staff officer to another until he was at last marched into Walker's office accompanied by the Chief of Staff and the ADC. This is what he had expected from the Army: 'red tape', pomposity and obstruction. So when Walker asked Hobbs what he wanted, he replied, 'I want to talk to you privately.'

When the others had left the room, Hobbs described the difficulties he had had in reaching Walker, adding, 'If this is the way you run your Army, no wonder men don't want to join it.' Walker was as annoyed as Hobbs had been at the difficulties and delays, of which he had known nothing; he would be delighted to talk about recruiting on television, he said, with or without a private soldier beside him. The interview was filmed and Walker was pleased at the effect of his first television performance.

Later, Hobbs came to lunch at Claxton Hall. Although he took every opportunity to say that he was no respecter of military rank or discipline, Walker had taken a liking to him. He was direct and blunt almost to the point of rudeness but he was shrewd and his approach was essentially constructive. While walking in the large garden, Hobbs looked at the well-kept flower-beds, the immaculate lawns and the freshly-painted house and remarked, 'My God, you live well. I'd like to make a film about the way you live.' Walker replied that he would be delighted if he did so and they agreed to think it over.

The Walkers did live well in Yorkshire, as the Army expected that they should. In the comfortable context of Claxton Hall, they were able to enjoy more of the relaxations of family life than had been usual. While here, Venetia became engaged to a young officer in the Royal Irish Rangers, Richard Venning, the son of one of her father's old friends from India, and they were married in York Minster. Nigel, who had also married, was with the 6th Gurkhas in Hong Kong but later resigned and returned to England, the third generation of Walkers to have served with the Gurkhas. Anthony, who had commanded a platoon of the Rifle Brigade in Borneo, had returned only to be seriously injured in a car accident; after six months in hospital he had recovered sufficiently to resume his Army career.

For Walter Walker, the future was now an open question. He had survived, vindicated, to be promoted and knighted and many of his friends considered that this, if not sufficient reward, was all that he could reasonably expect. Appointments for lieutenant-generals were, of course, even fewer than those for major-generals and this was particularly the case with Walker. It would have been surprising, for example, were he offered a place on the Army Board, which he had fought for so long, almost single-handed. In any case, he was universally recognised as a fighting soldier, a field commander, and, in peacetime, such qualifications put an officer in a long list for less than half a dozen jobs. There were only three such commands available for such officers in the British Army: the Strategic Command in Southern England, the British Army of the Rhine and the Far East, where future strategic plans were still fluid. All such appointments were filled by officers of the right age and seniority, making the final, carefully-planned steps in distinguished and unruffled careers.

There was one other appointment; one that might carry more direct and heavy responsibility than any of the others. At the end of 1968, Walker received another ominous envelope from the Military Secretary's Office. It contained a letter from General Goodwin giving the news that Walker was on a short list of officers being considered as the next Commander-in-Chief, Northern Europe, under the North Atlantic Treaty Organisation, and would he, if chosen, accept the appointment?

This possibility not only offered immense scope, three more years' service and promotion to full, 'four-star' General, but a chance to bring together and put into practice all the lessons he had learned and developed as a soldier. He felt certain he knew whom he had to thank for this chance: General Graf von Kielmansegg, who had shown such confidence in him in NATO and had reported as much to Whitehall.

A curious note in Goodwin's letter had been the manner in which he asked whether Walker would accept the job if offered, when it seemed obvious that he would be overjoyed to accept. This, it emerged, was because of one of the many sensitivities of NATO. Since Northern Europe Command had been formed, in 1951, to cover Norway and Denmark and, later, the Schleswig-Holstein province of West Germany, the Commander-in-Chief had been British. The first had been an admiral and his successors had been generals – at present General Sir Kenneth Darling – and, from time to time, the Scandinavians had suggested that they were capable of producing an equally effective Commander-in-Chief.

If Britain was to keep this appointment it was important that the chosen officer should like the Scandinavians and be liked in return. Perhaps Walker would miss polo or pig-sticking? It would be wonderful if he happened to enjoy winter sports and sailing. Amused, Walker's reaction was that he would be delighted to take up winter sports and sailing immediately and with much enthusiasm.

The prospect of commanding the vast and vulnerable northern flank of Western Europe and its intense contrast to all the active commands he had held hitherto was challenging and stimulating. He at once replied that he would, if chosen, be delighted to accept.

In February, 1969, it was announced in the *London Gazette* that General Sir Walter Walker had been appointed Commander-in-Chief, Northern Europe, and would relieve General Darling in August.

Reaction to the news in the Army tended to be delighted surprise or alarmed apprehension. Field-Marshal Sir Richard Hull, who had been succeeded as Chief of the Defence Staff by Marshal of the Royal Air Force Sir Charles Elworthy in 1967, told a friend that, had he still been in office, the appointment would have been made over his dead body. In NATO, those who remembered Walker at Fontainebleau and Brunssum were not surprised but, in Scandinavia, his name meant little or nothing, a state of affairs that would soon be changed.

At the beginning of the year, John Hobbs, the television producer, again approached Walker about the proposed programme, which, he said, he wanted to call *A Day in the Life of a General*. But there was now so much to do before the move to Allied Forces Northern Europe – AFNORTH – and so many farewell visits to be made, that Walker suggested that the plan might be postponed until he had taken up his new appointment. This would not only provide NATO with some useful publicity, which it needed, but, since he would be his own master and would not have to refer constantly to the Ministry of Defence for permission and guidance, the operation could be conducted without interference from outside.

At NATO Headquarters, Northern Europe, at Kolsas, near Oslo, there was one of the best information officers in the field of defence, Klaus Koren, who had had this job since AFNORTH had been formed.

In March, Walker made enquiries about the possibility of making such a film soon after his arrival and Koren was enthusiastic, suggesting that it could also cover a major NATO exercise due to take place in Denmark later in the year. All those, in NATO and Whitehall, who needed to know about the project were informed and signified agreement and pleasure at the prospect of free and, presumably, favourable publicity.

Shortly before Walker left York for Oslo, he visited London for meetings with Healey, Elworthy and Baker and also with Grandy, now Chief of the Air Staff and the brilliant Admiral Sir Michael Le Fanu, the First Sea Lord. They were friendly and wished him an enjoyable stay in Scandinavia but also gave a small warning. It had always been known that some Norwegians and Danes had questioned the automatic appointment of a British general to AFNORTH and had occasionally mentioned a candidate of their own. Just recently, the Norwegian Defence Minister had, in conversation with Elworthy, said that they would prefer a British admiral. He had been told that General Walker had already been chosen and that such a proposal was too late and that, in any case, such a request should be made not to the British but to the NATO Supreme Commander in Europe, the American General Andrew J. Goodpaster.

This incident was regarded as closed but was mentioned as an indication that the political background to the defence of Northern Europe might not be so placid as it might seem. To Walker, it seemed placid enough, when compared with South-East Asia, and, early in August, he flew to Oslo, ready to accept whatever challenges, military or political, awaited him.

The motive forces that had driven him thus far – through five campaigns in Asia, and to the top of his profession in the face of apparently implacable opposition from his superiors – showed no signs of slackening. What were these forces? Doubtless, a curious compound. There was something of the old Empire-builders' ambition, sharpened by family tradition and the strong, pioneering character of his mother. There was the competitive spirit fostered by English schools and by the Army. There was the conviction that when a right decision had been taken it had to be acted upon and carried through to its conclusion, whatever the opposition. All this had been tempered and sharpened on the battlefield.

As Walker prepared to take up his new appointment, he realised that his ruthless drive would still be needed for he was exchanging one sort of jungle for another.

8

The Nato Jungle

The choice of Sir Walter Walker to command the most vulnerable and sensitive flank of Western Europe caused intense speculation among the comparatively small number of senior officers who knew of his confrontation with the Army Board. Now, it seemed clear, he had not only recaptured his superiors' confidence and patronage but, since this particular tiger was unlikely to have changed his stripes, his appointment could only mark a change in NATO policy towards the Scandinavians.

Both flanks of the NATO line of battle were politically insecure and had needed careful handling. But whereas the southern flank periodically shuddered to seismic shocks from collisions between extreme right-wing and left-wing politics in Greece and Turkey and from conflict between the two over Cyprus, the northern flank trembled on the brink of neutralism. With the example of the comfortable fence-sitting of Sweden and with the contented, if emasculated, survival of Finland, the temptation to appease Russia by opting out of the West, or even leaning towards the East, was very real.

The shared experience of German occupation and liberation by the British and Americans gave the Norwegians and Danes a loyalty to the West, while the fear of the Soviet Union, which had shown what it could do to small nations by its destruction of the Baltic States, suggested that the sooner neutrality could be achieved the safer. As it was, Norway and Denmark had compromised and, while remaining full members of the Atlantic Alliance, refused to allow other NATO forces to be stationed on their territory – except for exercises – and, in peacetime, would have nothing to do with nuclear weapons or strategic missiles.

Hitherto, NATO policy had been to humour the Scandinavians for fear of upsetting this delicate balance of loyalties and apprehensions. Yet now a general had been chosen as their Commander-in-Chief who had never been known to humour anybody. Those who had known Walter Walker and his reputation could trace a consistent pattern as far back as 1944. The 8th Gurkhas in Burma and the 6th in Malaya had experienced this in its most direct form. It was the classic situation of military leadership, one that must have recurred since men first organised themselves into armies. A body of soldiers is drained of its dynamism – by too harsh a war or too soft a peace – and must be revitalised by a new commander. In doing this he becomes unpopular – even hated – but he succeeds and his soldiers realise that he has done so when they find themselves equal to the ordeals from which they had flinched.

Now, it was being said, Walter Walker had been chosen to give the Scandinavians the same treatment as those battalions, the same infusion of zest and confidence that he had given his brigade, his division, the forces in Borneo and, recently, to NATO forces in Central Europe. Denis Healey had already warned them at NATO meetings by saying that the new Commander-in-Chief was 'the best fighting general in the British Army'.

Although Walker himself had not been given specific instructions as to his conduct towards the Norwegians, Danes and the Germans in Schleswig-Holstein, he assumed that he was expected to behave as he always had. Indeed, it never occurred to him that anything else might be expected, because the British Army, at least, had had ample experience of his powers and techniques of leadership. They also knew of his definition of the soldier's responsibility. The stage and the scene must be set by the politicians whom he served. But, acting within these bounds, it was he who should be actor-manager. He had been chosen to prepare the defences of the northern flank to meet an attack from the East and this he would do by all means in his power. The politicians had given him a role and it was for him to perform it without unnecessary interference.

As General Sir Walter Walker flew north-east across the North Sea on 7th August, 1969, to take up his new command, he was troubled by no doubts over the task ahead. Instead, there was the excitement of a challenge. One was that his powers of leadership and command were to be tested to the full. The Norwegians and Danes had, through their politicians, asked to be defended against Russia; he had been chosen to lead them in this; now he would show them how it could be done. What he had seen of the Scandinavians he had liked, but it was obvious that they lacked some of the essential will and fire that would be necessary to face the potential dangers. It was up to him to give them this – just as he had given it to the 8th Gurkhas in the hills of Burma a quarter of a century before.

Another challenge was equally professional but more technical. Study of the map of his new command showed a remarkable parallel with his last responsibility in the field. While, climatically, Borneo and Norway could not be more different, there was a striking topographical and strategic similarity. In both there was an immensely long front to be defended – one thousand miles in Borneo, thirteen hundred from Hamburg to North Cape – and both commands had the sea at their backs. Here, as in Borneo, he would have to face potentially overwhelming odds. In both, there was the danger of the enemy within.

At Oslo, Walker and his wife were met by the Darlings. There was to be no long handing-over, for the transfer of power was to take place at midnight and General Darling would leave for London immediately.

Thoughtfully they had already moved out of their official residence, so the Walkers stayed at the large, wooden house at Holmenkollen, on a hillside outside Oslo, standing below the spectacular Olympic ski-jump and commanding a distant view of the fjord. The couples dined together that night and, afterwards, the two men discussed Walker's NATO inheritance. Darling seemed to have got on well with the Scandinavians but, Walker noted, he did say that it was necessary sometimes to 'jolly them along'. That would not be his way.

Next morning, the Darlings left for home and Walker found himself face to face with the day-to-day problems of defending West against East.

Should this need arise, Walker would, of course, have instantly taken command of all land, sea and air forces in the theatre; both those already in being and whatever reinforcements and support could reach him from outside. But, in peacetime, he was in direct control only of the headquarters and a skeleton command structure; the forces, which would be his in war, remaining under national control. Therefore it was a first essential to make contact and establish happy working relationships with the politicians and commanders of the three nations primarily involved in AFNORTH.

The first courtesy call in Norway would be to the Defence Minister, Otto Grieg Tidemand. A forceful man of forty-eight, Tidemand had served with the Royal Norwegian Air Force in Britain during the Second World War and had subsequently worked in his shipping business while nursing political ambitions. It had already become clear that he was out to make his mark by reforming the Norwegian defence structure and, perhaps, achieving the widely-shared aim of replacing a British NATO Commander-in-Chief with a Norwegian. In working towards the former goal, he had invited Canadian officers to Oslo to advise him on the possibilities of merging the land, sea and air arms into a single force. With an eye on the latter, he had a ready candidate, Admiral Folke Johannessen, who had been Chief of Defence for the past five years.

Walker knew that Tidemand had asked Elworthy that an admiral be given his appointment – a British admiral who might be succeeded by a Norwegian admiral – and he had been warned that the Defence

Minister was something of an autocrat. So, on meeting Tidemand, it came as no surprise that his first words of welcome were, 'You know, General, I gave orders that when you arrived you were to be put back on the aeroplane and sent back to England.' He said this with a smile but there was a bullying note in the banter.

He then welcomed Walker to Norway with some heartiness before reverting to the theme that he felt Norway had not been fully consulted in the choice of a Commander-in-Chief and that, when they were consulted before the appointment of a successor, they would insist on an admiral. Tidemand said that he had been told by Healey that Walker was 'a real professional' but had he not 'rocked the boat' over the Gurkha run-down proposal? Healey had told him so and added that he had been right.

Noting Tidemand's attempts to put him on the defensive, Walker took the initiative himself. 'Yes, events proved that I was right,' he said. 'And I think I am the only senior soldier who has rocked the boat like that and survived.' To press the point home, he added that he had managed to survive because he was not afraid of anybody, soldier or politician.

Beyond conversational sparring, Walker was not as yet able to discuss defence policy with Tidemand because he himself had not been briefed fully on the strategic and political situation in Northern Europe. This would take time and involve a tour of his command as well as conferences at his headquarters. These began soon after his arrival and he at once saw that the first of his reforms would be needed there.

Headquarters, Allied Forces Northern Europe, was at Kolsas, a small town a few miles to the north of Oslo. Here was not only the peace headquarters – modern office blocks in the neat Scandinavian style – but the war headquarters. The location of most NATO war headquarters, whether fixed or mobile, were secret but this, like some of the headquarters and strategic missile sites in the United States, was 'hardened' – virtually impregnable against even nuclear attack so that its position need not be regarded as secret. The war headquarters at Kolsas lay beneath at least a thousand feet of rock inside the mountain behind the peace headquarters.

The officers of the headquarters staff were of five nationalities. Just over half were British and American – because reinforcement in war would come primarily from their forces – and the remainder Norwegian and Danish and a few Germans. Although Walker had expressed his dislike of lengthy briefings, as soon as the first briefing officer arrived in his office it was obvious that they had been indoctrinated by the ponderous, verbose methods of the Americans. Each specialist had prepared to brief the new Commander-in-Chief by reading a lengthy paper, or by leaving it for him to read himself. In the American manner, this slowly built up to a conclusion, the reverse of the British method which was to begin a staff paper with its conclusion then explain concisely why that had been reached. Walker called the American-style briefings 'canned sermons' and would not listen to them.

Each relevant staff officer would, he ordered, visit him for fifteen minutes and in that time would tell him, without recourse to notes, what his job was and what his problems were now and might be in the future. If a staff officer could not do that, he said, he was not up to his job. All necessary briefings would, carried out like this, occupy only one day. Walker added that he wanted no formal introductions; he would walk round his headquarters and meet all his staff – officers, clerks, secretaries – at their desks.

There was resentment at Walker's arbitrary overthrowing of established methods and some staff officers assumed that he would relax his attitude. While they carried out his first briefings as he required, the written briefs that he would take with him on tour were drawn up in the traditional manner. They were returned to their authors, one with a note from the Commander-in-Chief saying, 'I found my brief of little value.... I want the briefs produced exactly to my wishes.... I trust all will realise that it is no good attempting to impose on me any other system.'

Walker's wishes were that briefs for tours should consist only of a map annotated with essential information. This would be an aide memoire rather than a brief but, as he intended to gather all necessary information himself from those directly concerned, no more would be necessary. Moreover, he announced, there would be no more formal staff briefings of visiting Very Important Persons. Once he had mastered the situation, he would brief these visitors himself.

For their part, the staff officers were surprised not only by the brusque manner of Walker's debut but by his personal ways and appearance. Darling had been, they thought, a typical Englishman but had obviously tried to make himself popular and had mixed freely and easily with his officers socially and in the bar before lunch. This was something quite different. One Norwegian officer recalled that at first Walker 'seemed grotesquely out of place – like something from the last century'. Details of his behaviour were whispered with almost feline inquisitiveness: that a red carpet had been laid outside his office, that his blotting paper was changed daily, that he sat in the front of his car with the driver so as not to crease his uniform. None of these rumours were, in fact, true.

Staff work apart, Walker had been struck by the curiously unmilitary atmosphere of the headquarters. Most of his officers worked what he called 'trade union hours' from eight in the morning until four in the afternoon, at that moment dropping whatever work had been engaging them and departing. Only the British and Germans seemed to have any sense of urgency and show readiness to stay at their desks until their work was done. He also noted that the Norwegians and Danes prided themselves on being 'democratic': there was no officers' club and the officers' mess was a cafeteria sharing the same cookhouse with the soldiers' cafeteria.

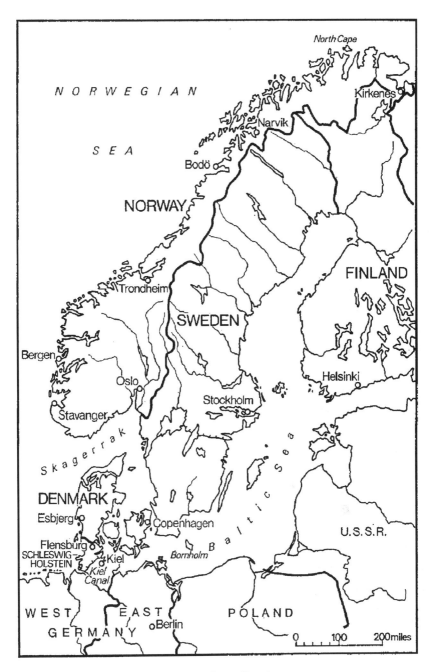

NATO in Scandinavia

On the credit side, he was impressed by the way Darling had organised the structure of the headquarters and, particularly, by the underground war headquarters, which was far more elaborate and efficiently arranged than he had thought it would be. He had yet to meet the commanders in the field and most of the politicians and civil servants in the capitals, so that his first impression was only of Kolsas. This was that the basic organism of AFNORTH was sound, but that its human heart was not. After his first weeks at Kolsas, Walker decided that only a quarter of his officers were up to the standard he expected.

Among the officers whom he found well up to standard were his Deputy, Lieutenant-General Wilhelm Mohr, of the Royal Norwegian Air Force, and his German Deputy Chief of Staff (Plans and Operations), Rear-Admiral Fritz Guggenberger, who, as a young U-boat captain, had sunk the Royal Navy's most famous aircraft carrier, the *Ark Royal*. In some NATO headquarters, the job of deputy was a useful means of keeping an ineffective officer, who could not be removed, out of the way without any risk of his taking offence. This was not to be the case with these two officers and Walker made this clear by moving Mohr into an office beside his own and telling his officers that he was Deputy Commander-in-Chief 'in every sense'.

Another member of the staff in whom Walker had great faith was Claus Koren, the chief of public relations. A Norwegian with an English wife, a long and intimate knowledge of NATO and a natural loyalty and honesty, made him a valuable adviser on public relations in its broadest sense.

One of Koren's tasks was to keep in touch with Tyne-Tees Television over the proposed film. At the end of August, John Hobbs arrived in Oslo to discuss the plans and possibly to change the approach of the film, now that it was to be made in a NATO and Scandinavian setting. The young producer appeared as brash in Norway as he had in Yorkshire.

To help put him at his ease in the unaccustomed surroundings of an international headquarters, Walker had had a memorandum circulated saying that, although he was young and inexperienced, Hobbs was a highly-regarded, professional producer and interviewer and should be accorded every assistance and respect. As was customary with visiting VIPs, a programme for his stay was drawn up and presented to him, typed, in a folder on arrival.

Hobbs looked at this and exploded, 'What's this bloody thing for? I've just arrived and I'm given a piece of bloody paper telling me what I'm going to have to do for the next three days. I'm not even in the bloody Army!'

But a curious relationship was established between the general and the television producer as mutual respect built up between them.

By Walker's standards, Hobbs's manners sometimes seemed questionable. Indeed, Koren was to tell him that anybody else 'would have found himself outside the gate very quickly indeed'.

And yet Hobbs stayed on and plans for the film went ahead. Walker had taken a liking to him because he saw that, behind the aggression, Hobbs was an acute professional and likely to achieve their common aim as quickly and efficiently as possible. He was also pleased to see that Hobbs was, of his own accord, not only coming to accept and even advocate the aims of NATO but was planning to make NATO, and not the personality of the Commander-in-Chief, the theme of his film.

For his part, Hobbs was surprised and impressed by Walker. 'He treated me like an equal,' he told Malcolm Campbell, his co-producer. 'He sticks to his word. I have the highest possible regard for him.'

Through Koren, it was arranged that filming should begin in September at Kolsas and later in Denmark during the amphibious Exercise Green Express. Both NATO Supreme Headquarters and the Ministry of Defence in London were pleased at what they regarded as a major public relations coup, for it was difficult enough to persuade television companies to make a half-hour programme about NATO, let alone one in which the producers turned out to be converts to the doctrine of the Alliance. The Danish and German Defence Ministries were informed – since they would be acting as hosts to the exercise – but the Norwegians were not, because all filming would be at NATO headquarters and their permission was not strictly necessary.

Hobbs and Campbell planned, with Walker's and Koren's agreement, that the film should take the form of an interview with the Commander-in-Chief about the aims of NATO and the Russian threat and that this would be illustrated by action sequences showing the exercise. As a dramatic introduction to the film, it was suggested that Walker be filmed arriving at his war headquarters and entering the main operations room.

This idea raised security problems. Both Hobbs and Campbell had been cleared to receive classified information but, obviously, care must be taken to avoid showing anything in the film that might be useful to Russian Intelligence. But there was no secret about the site of the war headquarters itself. A large, multi-level complex, it had been built by civilian contractors in the late Nineteen-Fifties and, had not its whereabouts been common knowledge in Norway, it advertised its presence by the communications aerials on top of the mountain. Three strongly-built entrances could be seen by anybody walking past the perimeter fence and, anyone interested enough to line up the bearings of these and assume that the headquarters would be under the highest part of the mountain, could make a reasonable guess at its position.

Photographs of the interior had been published in the *NATO Letter* magazine, but care would have to be taken that no accurate information was carelessly shown on any of the state boards in the operations rooms. So immense care was taken. The opening sequence of the film was to show Walker's staff car racing up a mountain road, but the road chosen was not to be one that led to the war headquarters. The Commander-in-Chief would be seen arriving at the main entrance and walking down the long tunnel through the rock into the heart of the mountain – but both these views had been photographed and published before. Finally, in the operations rooms all visible evidence of an actual strategic situation were to be removed or falsified. Nothing of value to the Russians would be seen.

In one shot, a teleprinter would be seen working, so, in case it might be possible to read the message on the television screen, its wording would read, '… and the cow jumped over the moon.'

To prevent any possibility of a security leakage, Walker decided that not only would each shot be carefully checked by his own intelligence staff, and by the experienced Koren, but that, whenever filming was in progress, he himself would be present. So, on 10th September, this began with Walker delighted with what he called 'a once-in-a-lifetime chance of allowing a "converted" television producer to go to his public with a hard-hitting story of NATO which might wake the public up and make people understand what NATO is all about'.

Filming at Kolsas and in Denmark was completed in less than three weeks and the Tyne-Tees team returned to Newcastle for the processing and editing, promising to show Walker the final version of the film in a few weeks' time. This was an essential preliminary to public showing because it had been agreed that, while Tyne-Tees would have complete editorial control, the film would have to be checked by NATO for any breach of security.

During October Walker had two brushes with Tidemand. One was over the failure of AFNORTH to fully inform the Norwegian Government of the plans for the film. This had been an oversight – despite the excuse that such consultation would have been pointless until it was known exactly how much filming in Norway would be necessary – but it had clearly caused affront to Tidemand. The second came when, at a party given by the British Ambassador, Tidemand suggested that Walker had spoken to the Press on some political matter. When asked to specify, Tidemand said that he had forgotten the details. This amounted to only a few sentences in a conversation, but it assumed the proportions of a warning of conflict to come.

Early in November, the Tyne-Tees film was complete and Hobbs flew to Oslo with a copy for Walker and the NATO staff to see. The first showing, on 17th November, was attended by Walker, several of his senior officers and Norwegian and Danish information and intelligence officials.

The film was as dramatic as had been expected but the audience watched for more than production technique and camera-work. Walker began by speaking about NATO, about the Soviet threat and the dangers of ignorance and complacency, just as many other senior officers and officials of the Alliance had been for years. Then he said, 'There is a great temptation to the Soviets towards aggression if we are complacent, if we drop our guard. We have got to do something about it, we've got to close this gap between what I as a soldier know to be the position on the ground and what the man in the street is told by the politicians and through the medium of his television screen. I feel very strongly that, over the years, NATO has built up sensible and reliable measures for dealing with any eventuality, any emergency and the unexpected – and the headquarters you are about to see is one of these preparations.'

Later, after Walker had spoken of the task of tracking Russian military movements, Hobbs asked whether NATO forces were being similarly watched. 'I am sure they are,' replied Walker. 'This is a game of dog eat dog.'

Did Walker think that Russia would ever deliberately wage war on NATO?

'I don't think she would embark on general war, because this would be mutual suicide. I think the danger is that there may be war by miscalculation, rather than premeditated war.'

What did he mean by miscalculation?

'Well, you may go so far to the brink that you spill over.'

Finally came the recurring emphasis upon the need to keep the NATO nations fully informed, Walker saying, 'How do you get across what we as soldiers know to be the position on the ground and the situation as given to the public through the medium of television, radio and the newspapers? This is the information battle.'

Hobbs then asked, 'But whose responsibility is it then to inform the public in Britain, let's say? The responsibility of NATO or the responsibility of the politician?'

'Both,' replied Walker. 'I think NATO's got a – could do a lot more – in waging this campaign for the dissemination of information about NATO.'

'Do you think the politicians could do more?'

'I think they could.'

'Then why don't they?'

'Well, perhaps it's up to people like me to persuade them to do so.'

When the lights went up, the audience agreed that it had been a stimulating film. Certainly there had been no infringement of security and, if Walker had spoken out bluntly about public ignorance and urged the politicians to take necessary action against it, then this was frank, slightly provocative talk that would make stimulating television instead of a bland propaganda film. This, it was agreed, was surely something at which only the most sensitive of thin-skinned politicians could take offence.

Two days later, most of the audience reassembled for the formal security vetting of the film by NATO that would decide whether any cuts would be necessary. They were joined by Major-General Ronald Coaker, the Chief of Intelligence and Security at Supreme Headquarters, Colonel Minton Wilson, the Chief of Public Information at SHAPE, and Brigadier John Woodrow, the British Army's Director of Public Relations. Coaker and Woodrow were British; Wilson, American.

When the film had been shown, the question of the information on the stateboards in the war headquarters was raised but once Coaker had been assured that it was all false he and the others agreed that there could be no possible security objections. Woodrow, however, made the point that there might be political repercussions to Walker's criticism of NATO public relations. But that, he added, was a matter for Walker's own discretion, not for theirs. It was decided that there was no reason why the film should not be shown as it was, without cuts of any kind. John Hobbs was not surprised but was disappointed that his film seemed so popular. He would have enjoyed some argument and a chance to stand up for freedom of expression.

Returning to Newcastle, Hobbs found that great interest had been aroused by the film, which was thought to be one of the best documentaries made by the company and there were hopes of selling it to Independent Television News for showing throughout the country by the other commercial companies. He took the film to London for showing at the ITN offices but, just before it was due to take place, an urgent message was received from the Ministry of Defence, saying that there were, after all, security objections to the film. After some argument it was agreed that it could be shown to a few executives of ITN and this was done. Although the defence correspondent, Peter Snow, thought that there was nothing new in Walker's opinions, it was agreed that this was vivid television and that ITN would like to show the film.

On 18th November, as the Tyne-Tees producers were wondering why NATO should suddenly change its mind, a telephone call was made to Walker's office from the Chief of Staff at SHAPE. He was told that although Coaker and Wilson had reported that there could be no objections, the Supreme Commander could not support their view. Although General Andrew J. Goodpaster had not seen the film himself, he had four objections to it. One was that some of the language was too flamboyant, notably Walker's use of the phrase 'dog eat dog'. There were other statements that could be considered controversial, or be misinterpreted. There must be no specific reference to the war headquarters and no view of its entrance must be shown.

That evening a signal arrived at Kolsas from General Goodpaster, formally asking that references to the war headquarters, and shots of its exterior, and controversial statements 'by military leaders', must be eliminated. It seemed likely that Brigadier Woodrow had reported fully on his doubts about the film on his return to the Ministry of Defence and that these, passed on to SHAPE, had started what seemed like panic.

Meanwhile, Healey had asked to see the film and a showing was arranged at the Ministry of Defence on the evening of the 20th, after Tyne-Tees, who owned the copyright, had agreed, but only on the understanding that this was a courtesy on their part and in no way accepted censorship. Hobbs and Campbell attended, together with Healey, John Groves, his Chief of Public Relations, General Baker, the CGS, and other senior officers and officials. After the viewing, Healey rose and said, 'Dramatic and compulsive viewing, gentlemen, but whether you get away with it is another matter.' Yet despite this ominous comment, Groves told the producers that there was no official British objection to their film.

Healey's position was ambiguous. He thought the film alarmist in that it suggested that a Russian attack was imminent. He considered its message to be so militaristic and insensitive to Scandinavian scruples that it might produce a fall in support for the Alliance. However, the film could only be stopped on security grounds.

Publicly and in the House of Commons, this was to be Healey's excuse. 'I was staggered that they showed the film of the entry road to the war headquarters and other things,' he told a journalist. But, privately, he admitted that the security objections were not strong. His fear was that Walker's warning was too strong meat for the Alliance to stomach. In NATO, as in Whitehall, the guiding rule was still to 'play it cool'.

He at once informed General Goodpaster of his misgivings, suggesting that he should see the film for himself.

Meanwhile, Walker had had to pass the Supreme Commander's objections on to Hobbs, who reacted angrily, for this was the sort of treatment he had always expected to receive from the military. The Ministry of Defence had suggested that he and Campbell take the film back to Oslo – the Ministry paying their fares – so that Walker could see it again and reconsider his original views.

At this third viewing, Walker's only criticism was that the film made NATO in general, and AFNORTH in particular, seem rather more efficient and effective than it was. But also in the audience was Kaare Sandegrem, the chief of the NATO foreign affairs department, and he announced that the Norwegians regarded the film as politically sensitive and that, in his view, it could only be shown on the understanding that it was not seen in Norway or Denmark.

Walker and his staff and Tyne-Tees were by now bemused by the dreamlike quality that suffused the affair. There seemed to be something about their film which aroused others to a frenzy of anger and apprehension, but which totally escaped them. The final touch of fantasy came when it was announced that Goodpaster, who had condemned the film without seeing it, now wanted to view it in Brussels. So the two producers, accompanied by Koren, flew to Brussels, nearly losing the film on the flight when the aircraft's baggage door fell open and an emergency landing had to be made.

The Brussels audience would have been worthy of a film première.

The Supreme Commander arrived by helicopter from SHAPE at Casteau, forty miles away, to join Dr Manlio Brosio, the Secretary-General of NATO, Sir Bernard Burrows, the British Ambassador, who was also the representative on the NATO Council, and about twenty other senior officers and officials. After the showing, several NATO public relations men were full of praise, one saying, 'It's a superb job – just what the NATO information services have been trying to do for years.' But Goodpaster strode out, without a word to the producers, passing Hobbs with, as he said, 'a face like thunder'.

Later, Brosio summoned them to his office.

Brosio came straight to the point. 'I hate that film,' he told them. 'It shows NATO as a warlike organisation. It is a political organisation – we are also ready to fight air pollution, for example.' It seemed that he was upset that there had been no reference to the political control of NATO commanders and by Walker's criticism of information services, for which he was responsible but felt could be adversely criticised only by himself.

At this point, negotiations foundered and sank and the producers flew back to London after some difficulty in getting hold of their film. Back in London, Hobbs kept the film under his bed in an hotel without protest from NATO, despite the fact that so much had been made of its dangerous contents and that Goodpaster had sent a telegram to the Chairman of the Independent Television Authority, insisting that it must at all times be kept under lock and key.

At Kolsas, Walker had been receiving and making long signals and telephone calls. The lengthiest of these were to and from Goodpaster and, on 24th November, orders arrived that 'it was essential that the following actions be taken if a breach of security is to be avoided' and there followed a list of mandatory cuts, including almost all the war headquarters sequence, which had already been seen and passed by the senior NATO security authorities. Any political problems connected with the film would be taken up by the Secretary-General direct with the producers.

Signals continued to fly between Kolsas, Brussels and London and, in Oslo, Tidemand complained that the film had been made without sufficient consultation with his Government, which also felt that it should not be publicly shown. Now, all the NATO and British military and political leaders were in opposition and Walker was told by the Supreme Commander that he himself must tell Tyne-Tees that, if the film was to be passed for security – and security was the only grounds on which NATO could take action – the listed cuts must be made. The buck had been passed back.

Walker had had enough. Throughout, he had acted in good faith but his superiors seemed to have no faith in him or his judgement. Nor did he feel that they were being honest with him, because he was convinced that there was no security problem but that NATO security was being used as a pretext for silencing his criticism of their political direction. Why could they not say so?

Now he was to be publicly humiliated by having to withdraw his earlier approval and tell Tyne-Tees to make the very changes that he had told them would be unnecessary. There seemed only one course open to him and that was resignation.

The issue was more than a conflict between levels in a politico-military hierarchy. It was one that had never been finally resolved and perhaps never would be. The issue was the role of the soldier in society. In civilised states, he was the servant of the people, with their political representatives as his direct masters. Yet was he more the servant of the politicians than of the people? If he were convinced that the political intermediaries were making wrong, even dangerously wrong, decisions, was it not his duty to speak over their heads to the people he served? Yet, if he did so, the politicians had the power to remove him and to put an end to whatever influence he might be able to exert through them by persuasion. Short of mounting a military *coup d'état* – and that would only convert a soldier into a politician – the dilemma was unavoidable.

But he was trained to obey orders, so, on 25th November, Walker ate his words as he had been instructed. He wrote to Lord Aylestone, Chairman of the Independent Television Authority, beginning, 'I regret to have to tell you that a considerable number of breaches of NATO security have now been discovered in the television production that Mr Hobbs and Mr Campbell were preparing', and he listed the Supreme Commander's original list of complaints.

He also sent a telegram asking that Hobbs bring the film back to Oslo as soon as possible 'to enable me to make at my level a complete security review of all photographic material and of the statements made by NATO military personnel pending a further review that may be required by higher authority'. He added that it was regretted that there were no funds available to pay travel costs.

For Tyne-Tees Television this was the last straw. Weeks of work and large sums of money had been devoted to the film. In reply, Arthur Clifford, who was in charge of current affairs programmes, issued a statement: 'The ruling that certain shots in the film might endanger NATO's security are, of course, accepted without question. But, as regards General Walker's own feelings on the need to keep the British public informed about NATO, Tyne-Tees can only suggest that in future he makes them known by other means.

'We are satisfied that Tyne-Tees has fulfilled its obligations to the letter. Quite simply, we have had enough. As far as we are concerned the Oslo file is closed.'

Their file was indeed closed. Tyne-Tees agreed to destroy all but one copy of the film and that they handed over to the Ministry of Defence, where it was locked in a box known as 'The Coffin' and placed in what was called 'the Cosmic Library vault'. But other files remained open, or were opened for the first time. The Press was now reporting the affair and Members of Parliament were alerted.

There might be Parliamentary questions, newspaper enquiries and prolonged rumblings of disquiet, but in NATO it seemed that the worst was over. General Goodpaster remarked that the affair of the film had caused him more work and trouble than any single problem since he became Supreme Commander. For Walker it had meant that he had rarely been able to leave his office before eight in the evening and, when he did, he carried with him at all times a bag containing the thick files of all the relevant signals. But what troubled him most was the embarrassing display of NATO floundering in a petty crisis. If they became as flustered as this over a simple matter of a television film, how would they behave in a real emergency? Was this an accurate example of their crisis-management?

Walker again considered resignation. On 16th December he wrote an appreciation of the situation and the reasons for his resignation. He also drafted signals to Goodpaster and Baker informing them of his decision.

Now he had second thoughts. It might be more effective to remain Commander-in-Chief and fight to change the attitudes he so detested. If he resigned, his place would not only be taken by an officer who would be unlikely to make such an effort but that officer might not be British, and Britain would lose a position of great value in political and military strategy within the Alliance.

The resignation signal was never sent. Two days later came a chance to fight. Goodpaster circulated a directive requiring military commanders to seek advice from their superiors before making any public statement that might reflect upon NATO policy. Walker's reply was blunt. Such a directive implied that four-star commanders, such as himself, were politically immature and, indeed, were to be treated like children.

Yet even as one crisis was ending another as beginning. Towards the end of 1969, Walker had been invited to lecture on his command and its problems to the Royal United Service Institution. A chance to address an audience of fellow-professionals in Whitehall had long been considered an honour among senior officers and past speakers had included most of the NATO Supreme Commanders and Commanders-in-Chief. Walker accepted and began to draft his lecture.

After the film affair, it would obviously be necessary to take extra care in avoiding anything that could be interpreted as an infringement of security, or a political embarrassment. But since Northern Europe Command was in more direct contact with the Russians than any other, it was impossible to speak about it without such references. The one hundred and fifteen miles of common border which Norway shared with the Soviet Union and the sixty miles of frontier between Schleswig-Holstein and East Germany could not be ignored, nor could the strategic situation at sea. Any school atlas would show that the routes to and from the Russian naval and mercantile bases in the Baltic and the Arctic ran through waters within the bounds of the command. It was equally obvious – had it not been frequently published – that NATO would keep a close watch by radar and other means on Russian sea and air activity in these areas.

Early in January, 1970, the RUSI advertised its forthcoming series of eight lectures culminating with 'Problems of the Defence of NATO's

Northern Flank' by General Walker on 25th March. To avoid any possibility of another conflict with his superiors, Walker informed the Supreme Commander, the Chief of the General Staff in London, the Norwegian, Danish and German Governments and the British and American Ambassadors in Oslo, what he was about to do.

First reaction was favourable. The Permanent Under-Secretary at the Norwegian Defence Ministry read the script and agreed that it was wholly acceptable. General Goodpaster was delighted, saying that the lecture would be 'useful and timely' and should enhance public understanding of NATO. General Baker was agreeable on condition that the script was first cleared by the Ministry of Defence and that Walker should be briefed in Whitehall before speaking. This did seem to confirm the advice Walker had received from a friend to the effect that the British Chiefs of Staff had lost faith in his political judgement. He also sent copies of the script to Goodpaster, Brosio and Tidemand personally, the latter being included as a courtesy since his Ministry had already approved.

Tidemand reacted sharply. The lecture itself was unacceptable. It was not that it infringed security but that its delivery by the Commander-in-Chief, Northern Europe, was most inadvisable. It made the Russians look like enemies and potential aggressors and Norway was trying to live in peace with Russia.

This touched off another explosion of signals and telephone calls almost on the scale of the film affair. Agreeing with Tidemand, Brosio now wanted the lecture stopped. But it had already been advertised and to cancel it would certainly look like censorship. Finally, all concerned agreed on a list of amendments or, rather, deletions. No mention must be made of Russian sea routes through the AFNORTH area, of surveillance of Russian activities, of the possibility of closing the Baltic in war or of Finnish relations with East and West. Emphasis must be laid on the defensive nature of NATO.

Rather than argue, Walker rewrote his lecture making sure that every statement of fact and every point he made had been published before, either officially by NATO, or in the Press, and not challenged by NATO.

On the afternoon of 25th March, the lecture-hall of the RUSI was crowded with officers and civil servants from the Ministry of Defence, diplomats and foreign defence attachés, defence correspondents and retired senior officers. But, almost at once, there was a sense of anticlimax, when it was realised that Walker had been gagged. He stressed that everything he was about to say had been said before and, throughout his lecture, referred to his sources.

Yet, despite what might have been crippling restrictions, the lecture had considerable impact. He described the Russian threat – quoting a speech by Tidemand in doing so – and how NATO was meeting it. He concluded with stress on the strategic importance of Scandinavia. If Norway and Denmark ever left NATO, the defence of the Continent and Britain would be seriously jeopardised and, with the northern flank turned, America's access to Europe would be exposed. All this was valid and delivered with panache, but it was obviously a lecture that had, as NATO jargon put it, been 'sanitized'.

Then came the questions. One of these was about the importance of training NATO forces from outside AFNORTH in Norway. The Norwegian defence attaché to London noted that Walker said in reply that 'off the record I can tell you that at any given moment there are two thousand Allied servicemen training or exercising in Norway'. Later, as the meeting broke up, the Soviet defence attaché, General Nemchenko, approached the Norwegian and accused his government of hypocrisy. The Norwegians had always claimed that foreign troops were never stationed in their country except for brief exercise periods – but now the Commander-in-Chief himself admitted that there were never less than two thousand!

Yet another explosion followed. Tidemand was furious. Walker protested that he had been misquoted, that he had said that that number trained or exercised in the Command in any one year.

Relations between the two were about to deteriorate further. Unlike the troubles over the film and the lecture, this issue did not threaten repercussions outside the restricted circle of defence planners, commanders and the politicians concerned. An abstruse question of command and control methods and their relative value, it nevertheless came near to bringing about a crisis in NATO relations.

The Norwegian armed forces were small, and their standing in Norwegian society was not high, but they constantly commanded the earnest attention of their country's legislators and commentators. From time to time parliamentary commissions studied defence problems and the most recent of these had been studying types of command structure.

The land defence of Norway was based on two regions, north and south, but only in the north was there a force in being. This was a brigade group and it was planned that, in due course, another would be formed to cover this southern part of the country.

Walker fully approved of the proposed command structure which would be much like that in Borneo. The Task Force Commander would be free to operate as Director of Operations giving his orders to his three service commanders who, in turn, would be free to use their own authority and expertise. This, in fact, would be the method he would use should an emergency give him command of all Norwegian forces. It was also the system favoured by General Goodpaster.

But neither Tidemand or the Norwegian Chief of Defence, Admiral Johannessen, agreed with this plan. They wanted a streamlined, unified command structure with the Task Force Commander in complete control of all three arms and with his senior land, sea and air officers as no more than advisers.

It seemed to them that a compromise should be possible. The land, sea and air commanders could be appointed as such, but in name only, since the Task Force Commander could use them as advisers and the streamlined structure could be preserved. When Walker heard of this he reported what he considered to be double-dealing to the NATO Supreme Commanders in Europe and the Atlantic. He was also angry at the suggestion, which was being put about by the Norwegian Ministry of Defence, that he himself had changed his mind and now agreed with Tidemand's compromise.

A breakdown in relations between NATO and the Norwegian Government might, it seemed, be avoided by the imminent replacement of Tidemand by the younger, more amenable Mr Gunnen Hellesen. But, in handing over, Tidemand would accompany his successor to a NATO Council meeting in Venice. There he fired his last shot.

Meeting General Goodpaster, the Norwegians found that the Supreme Commander was anxious to know exactly what claims they had been making in pressing their plans, perhaps embarrassingly anxious to know. When Goodpaster was told that Norway could no longer tolerate Walker as Commander-in-Chief, he proffered no sympathy. Tidemand left the meeting so hurriedly that he left his hat behind.

When the Norwegian Defence Committee considered the command and control proposals later that month, they concluded that the command structure should be that which Walker had always insisted was essential.

These first ten months as a commander of the defence of the West had been bizarre and wearing and had involved an extraordinary waste of time and effort. Walker had stood in danger of dismissal and always he had seemed to stand on a quicksand instead of a firm military and political base. Looking back over the sequence of misunderstandings, rages, face-savings, embarrassments, tantrums and confusion, it was hard to attach any significant blame to him. In the affair of the film, he had been mildly – although perhaps tactlessly – critical of politicians' public relations. In the affair of the lecture, he had behaved with punctilio. In the affair of the command structure, he had behaved like a forthright soldier. He had been fighting what he called 'the Information Battle'.

The greatest support of all had been Beryl. Visitors to their house often considered her the ideal soldier's wife. She was always calm and her judgements came from the head as well as the heart. Always beautiful, elegant and charming, she was the perfect foil to the soldier who believed that his profession should be taken as seriously in peace as in war.

The Walkers' house outside Oslo and their Copenhagen flat always seemed oases of sanity in a temperamental world. This was helped by the imaginative idea of replacing the British soldiers on the domestic staff by Gurkhas. Predictably, this caused a flutter in the Adjutant-General's office because it had never been done before, there were no regulations governing such a procedure and, probably, because it might smack of colonialism in liberal Scandinavia. But the Gurkhas remained to win admiration for their efforts in winter sports, as for their smartness and efficiency. At formal dinner parties, a Gurkha pipe-major performed the deafening ceremonial with the bagpipes to such effect that the United States Ambassador – Philip Crowe, who proved to be as helpful a friend to Walker as Stebbins had been in Katmandu – would invite him to play at his own functions. When, in Oslo, NATO officers were threatened in their homes by an anarchist group, the principal precaution taken by Walker was to tell his Gurkhas to be seen sharpening their kukris in the garden.

A heartening experience now followed the months of worry and humiliation, one that would demonstrate that there were still – as Walker would put it – sahibs as well as babus. This was a return to East of Suez, principally as Colonel of the 7th Gurkhas on an official visit to them in Hong Kong and also to tour Malaysia, Singapore, Brunei and India. Everywhere he was received with delight and accorded privileges and confidences that now rarely came the way of an Englishman. Indeed, the various British High Commissioners seemed to look upon these honours and this trust with envy. In Kuala Lumpur, he wrote to a friend, that he was 'made to feel I was one of them and that my Malaysian knighthood was not a mere courtesy title'. In Brunei, the Sultan invited him to return as defence adviser when he retired from the Army.

It was in New Delhi that he achieved, without trying, what could only be regarded as a diplomatic coup. Relations between India and Britain had been strained and now it seemed that Russia had become India's closest ally, particularly in defence co-operation. In this field there was little or no contact with Britain.

The British High Commission in New Delhi was therefore astonished to hear that Walker was staying with his old friend General Sam Manekshaw, who was now Commander-in-Chief of the Indian Army. The two soldiers spoke frankly and with complete trust. Walker particularly noted that Manekshaw's attitude to politics was not unlike his own. Politicians had their job and soldiers had theirs and the two should never mix. Indeed, not only did Manekshaw stress that the discipline and efficiency of the Indian Army depended upon its being totally divorced from politics – he rejected all thought of military rule in India – but that he himself made efforts to avoid personal contact, even socially, with politicians. Walker wished that he could have done this in Oslo.

Manekshaw was Colonel of the 8th Gurkhas, Walker's family regiment, which was now in the Indian Army, and officers of the regiment, past and present, and their wives, entertained Walker to a formal luncheon. 'They treated me as if I were their Commanding Officer,' he said afterwards. 'And made me feel as if I still belonged to the regiment.' This meeting confirmed his earlier impressions that Gurkha regiments in the Indian Army were as well led and as effective as they had been under British command and he noted with pleasure that all the regimental traditions and customs were lovingly preserved.

Walker returned to Europe, refreshed and stimulated, old ties strengthened.

During a Commander-in-Chief's first year, he was expected to have toured his Command, met his subordinate commanders, inspected his forces and formed his own assessments of the strategic situation. In strategic importance, his command was second in importance only to Central Europe but, since any Russian aggression in the centre would undoubtedly lead to a rapidly escalating 'nuclear exchange' – as the horrors of mutual nuclear bombardment were laconically known – it seemed more likely that attempts would be made against the flanks. The southern flank – Turkey, Greece, Italy and the Mediterranean – was obviously of great importance since it commanded the access to Middle East oil and to Africa. But the northern flank was of far greater immediate importance. This flank was the thin shield that stood between Russia and her Warsaw Pact allies and the British Isles and the sea routes linking Western Europe with North America. Since the basis of NATO defence was the support of Europe by North America, the importance of AFNORTH needed no further explanation.

The front stretched from the river Elbe in Germany to the town of Kirkenes in Finmark and beyond to the Arctic ice. It was split by the mouth of the Baltic and complicated by its outpost on the Danish island of Bornholm to the east of Denmark. One quarter of the eastern boundary lay against Russia or East Germany and the remainder against neutral Sweden and Finland, the latter having been at the mercy of the Soviet Union for the past thirty years. The value of this barrier on land and sea to the Russians was not only that it commanded access to their two major maritime bases in the Western Hemisphere – around Leningrad and Murmansk – but that, as the Germans had found in the Second World War, a pre-condition of aggression beyond the Continent was the control of Norway. The maze of fjords and islands facing the Atlantic and Arctic gave Norway a coastline of two thousand five hundred miles and innumerable deep-water harbours, most of them ice-free in winter, and it was from these that the Germans could have isolated Britain from America during the Second World War had the development of their surface fleet been more advanced.

The scale of this front was most vividly illustrated by the variety of its terrain and climate. In Schleswig-Holstein and Denmark, the country was gently rolling farmland well served with roads and railways, offering no obstacles to an invader, beyond the shallow Baltic waters between the Danish islands and the Kiel Canal across the Jutland peninsula. Across the Skagerrack channel, the bulge of southern Norway was mountainous but, as the Germans had found, offered no impregnable barriers to conventional armies. The long, narrow strip of mountains and fjords north of the great natural harbour of Trondheim was something quite different. The railway ended at Bodo, less than half-way to the extreme north, and the single road was often unmetalled and blocked by snow and ice in winter and broken by ferry-crossings. In the Arctic regions – almost totally dark throughout the winter – the mountains and the high plateau of Finmark were as wild and roadless as Borneo, but here the temperature often fell below minus forty degrees Centigrade and the weather could be as harsh as anywhere in the world.

There was variety among the people. Culture, temperament and recent history had set the Germans apart from the Danes and Norwegians, who considered themselves, together with the Swedes, a separate Nordic race. Less than five million Danes, in about seventeen thousand square miles, were principally occupied with farming and had not been known for centuries as a martial race, indeed there were foreigners living in Denmark who unkindly considered them a nation of pleasure-seekers who would not fight to defend themselves. Less than four million Norwegians occupied their one hundred and twenty-five thousand square miles, making Norway the most sparsely-populated country in Europe. Only three per cent of the land could be cultivated and the Norwegians relied for their prosperity upon their five million tons of merchant shipping, and fishing the seas. They were a stubborn, if somewhat smug, people and, as they had shown during the German occupation, capable of putting up a dogged defence when aroused.

That Norway and Denmark formed a tempting strategic prize for Russia was not in question. But while opinion in the West, lulled by more than twenty years of peace under NATO protection, seemed generally to doubt whether the Russians had any active designs on Scandinavia, to Walker it was an article of faith that they would risk anything short of nuclear war to neutralise these countries as an essential preliminary to westward expansion. When these Russian intentions were doubted, he would quote belligerent speeches by their military leaders. One of these, General Suvechin, he would quote as saying that the role of the Soviet Army was 'to stand by, ready to shake the tree when the rotten fruit is ripe to fall'. The aims of the Soviet Fleet seemed more active and Walker would quote the Russian Chief of Naval Staff as defining them thus: 'In the past, our ships and naval units have operated principally near our coasts. Now we intend to be prepared for broad offensive policies against the sea and land forces of the imperialists in any part of the world's oceans and adjacent territories.' These and other quotations left him in no doubt that the policy of the Soviet Union was dedicated to the reduction of the West.

When asked how the Russians could achieve this, Walker would remark that they were the finest chess-players in the world. They could outflank Norway at sea and, without resort to force, demonstrate to the Norwegians that, as they could expect no help from outside and no mercy from the Russians, it would be prudent to opt out of NATO and into neutrality. There was even a risk of territorial aggression on a limited scale, sudden, unexpected and supported by a manufactured political excuse to bemuse NATO and world opinion.

Remote Finmark was particularly threatened in this way. Soviet propaganda harped upon the use of the province as a NATO training area and, as such, a threat to Russian shipping and the freedom of the seas. A fabricated crisis, a *coup de main* by airborne and amphibious forces to seize Kirkenes, the radar installations and the airfields, followed by an appeal to the United Nations to halt NATO aggression, and what could the West do? Would the President of the United States resort to intercontinental nuclear war to recapture the barren mountains of Arctic Finmark?

At the opposite extremity of the Command there was the island of Bornholm, which not only lay east of the longitude of the Polish border and the 'Iron Curtain' but had been occupied by the Russians for ten months after the Second World War. Would the West be willing to risk nuclear destruction to save Bornholm?

The Russian forces available for immediate use in such adventures, or for more subtle, long-term demonstrations, were formidable. North Norway could be threatened by the strongest of the four Soviet fleets, including at least one hundred and sixty submarines – more than fifty of them nuclear-powered – and some two hundred and fifty surface warships, about thirty of them armed with missiles. On land, there were two mechanised divisions in the Kola Peninsula and five more in immediate reserve. Recently amphibious forces had been formed and a brigade of marines with assault vessels capable of carrying two thousand five hundred men were being trained in beach-landing exercises when the defenders were always referred to as the enemy.

In the Baltic, Warsaw Pact naval forces included forty submarines, about thirty cruisers and destroyers, nine of them missile-armed, and one hundred and seventy frigates and escorts. In addition, there were some two hundred and fifty fast patrol boats, a third of them armed with missiles. On land, there were ten Warsaw Pact divisions, of which three Russian and two East German directly faced Schleswig-Holstein. Amphibious capabilities were even greater than in the Arctic with two Russian brigades of marines, a Polish amphibious division, an East German brigade and assault craft to move them all. Shipping was available to lift a follow-up force of five divisions.

Within striking distance of AFNORTH were three airborne divisions, and the aircraft to lift two of them to any part of the Command without flying over neutral territory. Not counting these aircraft, Warsaw Pact air forces stationed within striking distance numbered at least two thousand three hundred military aircraft of all types. Almost all potential targets in the command were within range of more than six hundred medium-range missiles with nuclear warheads, not to take account of about fifty missile-firing submarines of the Soviet Fleet, many of them on station in the North Atlantic.

Behind this array was, of course, the main might of Russia: two million soldiers, the second largest navy in the world, more than ten thousand combat aircraft and thirteen hundred inter-continental nuclear missiles – and this only the strength immediately available.

The NATO forces Walker had to prepare to meet this gigantic threat were, by comparison, ridiculously puny. There was only one division in being and that was German – equipped with tactical nuclear weapons under American control – and was for the defence of Schleswig-Holstein. Like all NATO armies but the British, the German was based on conscripts and these were required to serve eighteen months, the minimum considered essential by the Supreme Commander. The Danes maintained four infantry brigades – mostly conscripts serving only a year – but these could be expanded, by the calling up of reservists, by three more brigades and fifteen local defence battalions. There were, in addition, nearly eighty thousand Home Guards. Norway's peacetime army – also based on conscripts serving a year – amounted to only one brigade stationed at Tromso in the north and an independent battalion. Its peacetime establishment was twenty thousand but this could be multiplied by six on mobilisation, and seventy thousand Home Guards were on call.

At sea, the whole of the German Navy – a useful force including seventeen destroyers and frigates, eleven small submarines and dozens of coastal craft – was assigned to AFNORTH. The Danes manned six frigates, six submarines and a fleet of coastal craft. The Norwegians had five frigates, fifteen small submarines and coastal craft.

In the air, the Germans, Danes and Norwegians each contributed just over one hundred combat aircraft each.

This manpower seemed small enough on paper but, in reality, it was even weaker; by a percentage between seventeen and thirty-five per cent. Even then, a quarter of what was available were recruits.

As part of NATO, these forces would not, of course, stand alone. But, since the Norwegians and Danes refused to allow foreign forces to be stationed on their territory, these would have to travel across the

North Sea or the Atlantic after the alarm had been sounded – and then it might be too late. Immediate reinforcements available included – should circumstances allow – a division and a parachute brigade from Britain, a brigade from Canada and, most important of all, the Ace (Allied Command Europe) Mobile Force. This was a brigade formed of one battalion from each of three NATO countries, its presence on a battlefield committing the whole Alliance. Air and naval forces would also be available and the core of this, indeed of the whole reinforcement operation, would be the Allied Striking Fleet, built round the strategic and tactical air power of its aircraft carriers and its division of marines.

This sounded formidable, but could it reach the scene of crisis in time? During the last Arctic exercise, Supreme Headquarters had been pleased that the Ace Mobile Force – including battalions from Canada, Italy and Britain – had been deployed in North Norway within seven days.

Since he commanded no forces in peacetime, Walker could only plan, advise, cajole and encourage those that he might have to lead. While there was no question of his command being able to defend itself against Russian attack on its own, this could be done if it could be fitted tightly into the NATO military machine and the gears tuned to the right pitch. Time was the vital factor; time to warn, time to react. Above all, time was needed for governments to consider the consequences of their acts.

The first necessity was for Allied Forces Northern Europe, to hold out long enough for reinforcements to arrive. Failing that, they must at least be able to put up a defence that would raise the threshold of commitment and force the aggressors to use such force as could not be described as 'police action', or excused as an isolated incident. They must have a straight choice of war or peace.

When Walker attended the exercises, which followed one another in wide variety and rapid sequence throughout his command, he formed plans for this defence. He had always been receptive to ideas from others, and some of these came from the Royal Marines, whom he had come to admire increasingly since the early days in Borneo. In one exercise for the Ace Mobile Force in North Norway, 45 Commando, which specialised in Arctic warfare, acted as the enemy and, in war, they would have won.

Two incidents particularly impressed Walker, who was living in the field among the troops. One was when the commandos captured a Norwegian fortress. They did this by discovering the time that the garrison's breakfast was served. Then, having stealthily occupied commanding features nearby, one troop embarked in helicopters, which flew low, round the contours of the mountains – just as they had in Borneo – and hovered over the fortress, while the commandos slid down ropes on to its roof and threw smoke bombs through the air vents.

The other was when an Italian and a Canadian battalion were advancing across snowfields into the hills but had not made contact with the enemy. The dominating feature of the terrain was a ridge between the two battalions, from which an enemy could dominate both. The Italians were slow in their advance – enjoying their coffee, it was said – and it was they who were due to occupy the strategic ridge. Walker, at the Canadian command post, asked why this had not been done or, for that matter, why the Canadians had not seized it themselves. The Canadian colonel replied that this was the Italians' task and they might be offended if he interfered. Both were too late, because 45 Commando not only took the ridge themselves but sent a raiding party on skis down from the summit to plant dummy bombs in the Canadian transport park. This incident impressed Walker, perhaps because, unconsciously at first, it reminded him of another ridge, more than a quarter of a century before, when this same situation had arisen at the Battle of Taungdaw.

The fiasco at the fort had shown how easily massed Russian airborne troops – let alone a few score commandos – could 'take out' fortified positions. Such garrisons – guarding airfields, radar and other fixed installations – must not only be alert but must be able to put up the necessary defence. He therefore advocated tactics around a system of 'porcupines' in vital forward areas, positions reminiscent of the defensive 'boxes' in Burma and the jungle forts in Borneo. Each would be held by a battalion group with its own artillery, tanks, anti-aircraft guns and helicopters; strong enough to hold out against attack by anything less than a division. If it held, there would be time to rein-force, If a division were thrown against it, the Russians would have the inevitable choice between war and peace. Thus the 'porcupines' in themselves would constitute a deterrent.

The incident of the ridge and the raid reminded Walker of cross-border raids in Borneo and how these had thrown superior Indonesian forces off balance, just as they had the Ace Mobile Force. If the AFNORTH armies were to defend 'porcupines', they could not find the manpower to mount raiding and guerrilla operations as well. But one of the strengths of both Norway and Denmark was in its reservists and Home Guards. Instead of these being used defensively in guarding bridges, supply dumps and other sensitive points, which they could not hope to defend against determined attack, they could be used offensively as guerrillas. Walker visualised six-man teams – armed and trained as tank-hunters or saboteurs and able to operate as inde-pendently as had the SAS in Borneo – as an integral part of his first-line defences. The Russians would know that, even if they destroyed the 'porcupines', they had yet to take the country itself and that they faced a long and costly counter-guerrilla campaign that would make a *coup de main* and a *fait accompli* an impossibility.

But the best-laid defence plans would come to nothing without one subtle ingredient that could never be entered in a defence bud-get: the will to fight. Always, at the back of Walker's mind was the question of what the Norwegians, Danes and Germans would be willing to sacrifice for their freedom. Many forces were attempting to erode their fortitude. Soviet propaganda was aimed specifically at isolating them mentally from the West – as Russian forces were trying to isolate them physically – and to subvert the idealism of the young.

That this was succeeding seemed to be borne out by the rise in conscientious objectors in all three countries. In Denmark, the total multiplied fifteen times in the ten years preceding 1970 and, if it continued to rise at that rate, there would not be enough conscripts to meet basic military obligations by 1973. In Norway the number had more than tripled in the same period and was expected to rise. In West Germany, nineteen thousand conscientious objectors were registered in 1970.

In September, 1971, the Danish coalition government resigned and the Social Democrats came to power. Their defence programme included the cutting of conscription service to five months, maintaining a standing army of only seven thousand and cutting the navy and air force by half. This would, if implemented in full, not only deprive Denmark of viable defences but would force the Norwegians to redeploy much of their forces to cover the south – which could be held lightly while the Danes fulfilled their obligations – as well as the north.

In Norway and Germany the danger was not so acute but, in both, the cutting of defence expenditure remained an attractive political attitude.

There was only one form of action that Walker could take to help strengthen resolve and that was to exercise his powers of personal leadership. He must both warn these people of the peril in which they stood and encourage them to believe that they could save themselves if they had the will. He decided that he himself would brief – in the form of lectures, speeches, interviews or conversation – anyone who would listen.

Not only could Walker talk openly about the threat and the possibility of a Russian attack but he would give full details of the Russian offensive capability and then ask why the Soviet Union should devote such massive effort to this if it was not to be used.

He became something of a celebrity, an intriguing mixture of traditional British officer and a provocative public personality. 'Outwardly,' one journalist wrote, 'he could have been a pattern for the projection of Alec Guinness of the uncompromising Indian Army officer in the film *The Bridge Over the River Kwai* – only with the difference that General Walker is a modern officer with a sense of public responsibility.' A German reporter wrote, 'General Walker comes very close to the prototype of the British Colonial officer, with his grey moustache, the cane under his arm and the polka-dot handkerchief which he knows how to produce gravitationally from his left sleeve.'

At the end of 1970, while in England for the funeral of Lord Slim, Walker was told that he was to be succeeded early in 1971 by Lieutenant-General Sir Tom Pearson, who had followed Goodwin as Military Secretary. He was given the news in confidence – finding that it had leaked at a cocktail party attended by Army Board wives, just as his own appointment had three years before – but the appointment had also been agreed to by AFNORTH governments. This meant that Norwegian hopes of replacing a British general as Commander-in-Chief had failed. It also meant the end of any plans there may have been in this direction for Admiral Johannessen, for his own days as Chief of Defence were now numbered.

As the end of his own time as Commander-in-Chief approached, Walker became increasingly convinced that his most urgent task was to continue and intensify his public relations activities. One of many interviews arranged by Koren was one at Kolsas on 25th October, 1971, with a Dutch television reporter. Walker gave his usual briefing, there was a general discussion and he was then interviewed before the camera. Also present was a reporter from the Amsterdam newspaper *De Telegraaf*, but he took little part in the discussion and seemed to be present as a spectator. The occasion followed the usual pattern and the discussion period was customarily robust, since it was not intended for quotation.

On 9th November, Walker was told by Supreme Headquarters that *De Telegraaf* had published an article which might cause trouble. When he read the translation, he saw that while most of it reflected his own views in the eye-catching way he hoped such articles would, several passages would catch and offend too many eyes.

It began, 'The British General Sir Walter Walker, Commander-in-Chief, Allied Forces, Northern Europe, at his desk in the Headquarters just outside Oslo, casually puts it this way: the only things his troops have and the Warsaw Pact forces have not is long-haired soldiers....' And again, 'Denmark worries me considerably.... I expect people there assume that long hair frightens the Russians.' Of the Norwegian and Danish policy to exclude foreign bases, he was quoted as saying, 'They don't know what they are talking about. Just take Norway. Now, in wartime this country is hopelessly lost....'

Walker remembered making jocular remarks about long hair during the off-the-cuff discussion but the final quotation was taken out of its context and given another meaning. He had, in fact, said that those who dismissed warnings of the Russian threat as alarmist 'don't know what they are talking about'. And he had stressed that Norway would be hopelessly lost unless reinforcements could arrive in time.

This was a tiresome confusion but of a sort familiar to those who are frequently interviewed. But this time, it seemed, no explosion had been triggered in Scandinavia, although Admiral Johannessen was said to have discussed the incident in the Ministry of Defence and Mr Alv Jakob Fostervoll, who had succeeded Hellesen as Minister, had been informed.

Walker regarded the incident as closed. On 20th November, the Walkers flew to Brunei as guests of the Sultan for the unveiling of his memorial to Sir Winston Churchill and there was the refreshing experience of lunch with officers of the 6th Gurkhas, who were guarding the Seria oilfield. A few days after his return to Oslo, the time-bomb, which had been ticking unnoticed, burst.

Three days before the official visit of the Soviet Prime Minister, Mr Kosygin, to Norway, the Oslo radical newspaper *Dagbladet* published the *De Telegraaf* article under the headlines, 'Kolsas Chief Attacks Norwegian Defence Policy. Norway Would Be Hopelessly Lost In

War'. It was now an instant sensation throughout Scandinavia and, that same day, Walker received his first rebuke. This came from the new Defence Minister, who said in public, 'It is unfortunate that General Walker is not better able to control how the newspapers understand his statements....' Fostervoll added that he had been told that General Walker ended his NATO assignment on 1st February, 1972, and that he did not intend to comment upon this at present. On television that evening, he was less reticent, saying that he regarded the affair as 'unfortunate, even deplorable'.

The acting Foreign Minister, Mr Thorvald Stoltenberg, went further with, 'I am glad he is leaving soon – this cannot be tolerated.'

For a fortnight, the Walker affair seemed to dominate the news in Scandinavia. At last the Scandinavians could feel themselves the centre of attention, it seemed to some diplomats, and could at the same time indulge themselves with a sense of outraged propriety and self-righteousness. Some could denounce Walker beneath headlines like, 'The Sword-Rattling General' and 'The Talkative General' yet a surprising number came to his support with 'General Walker's Unpleasant Truths' and 'Do Not Disturb the Ostrich, General!'

Trenchant support came from a Norwegian military magazine *Offisierbladet*. A leading article questioned the relative responsibilities of soldiers and politicians. 'After the incident with General Walker it is natural to raise the question whether a military commander – within certain security limits – must have the same right and duty to make his view and evaluations known in line with other professional leaders of our society. That this might not suit our politicians and others very well is no argument.

'Or are generals only experts and advisers to the responsible politicians? In such case, who is to inform the common man in our democratic society how the military professionals look upon the situation?'

At the NATO Council meeting in Brussels, early in December, Fostervoll and Admiral Johannessen at once set about complaining bitterly to General Goodpaster and to the British Minister of Defence and the Chief of the Defence Staff, Lord Carrington and Admiral of the Fleet Sir Peter Hill-Norton.

Goodpaster tried to smooth the ruffled feathers. But, while Walker now had the Supreme Commander's support, the British Chiefs of Staff Committee were embarrassed and annoyed.

It now became known at Kolsas that an issue at the Brussels confrontation had been the possibility of dismissing Walker. Both Norwegians and Danes had asked for his head, but had been dissuaded on the grounds that it would be disruptive for the Alliance and that, anyway, he was due to leave in two months' time.

Sympathisers, expecting to find Walker downcast, were surprised to discover him buoyant. This, he was saying, was exactly what he had tried to achieve with the Tyne-Tees film and had been trying to achieve for the two and a half years since. His object had been to make people aware of a danger. This he had done and that he had done it at the cost of what might appear to be a massive breach of military etiquette worried him not at all.

It surprised some in NATO that the Commander-in-Chief had been belaboured over a petty misunderstanding with a reporter when he had been airing far more controversial theories. Some of these were offered to a defence studies seminar organised by the United Kingdom Western Command at Birmingham University in the autumn of 1971. Invited to forecast events of the next decade, Walker asked his audience to imagine they were looking back from the Nineteen-Eighties. The Soviet system had remained ideologically and politically committed to the destruction of the West. Their methods changed, but never their aim.

'By alternating pressure with relaxation, the Russians achieved their aim without firing a shot,' he warned. 'No wonder Krushchev visualised leading us to our grave with one arm round our shoulder.... We ignored what Trotsky said in 1929, when lecturing in Copenhagen, that the world revolution would start in Greece and that the next country would be Norway. This showed the Soviet concept of activity on two flanks.'

So, starting with his command, he said, as if composing his epitaph, 'In 1971, General Walker, who was Commander-in-Chief, Allied Forces, Northern Europe, drew attention to the colossal Soviet naval build-up based on Murmansk. This, in his view, could only have a sinister purpose. He was convinced that it was their intention to push their defence line outwards to Iceland and the Faroes and to turn the Norwegian Sea into a Soviet lake, behind which, of course, Norway would lie. Thus his command would become outflanked and isolated....

'This was only the first stage in further Soviet naval expansion into the Atlantic. Their aim was to put up a barrier at sea, between the United States and Europe, thus placing Western Europe in the same position as General Walker's command was to find itself....

'By the Nineteen-Eighties, the Soviets had reached a stage at which the sheer disparity of military strength left NATO's northern flank with no convincing strategy. Political pressure, backed by the threat of this greatly superior physical force, compelled Norway to lapse into neutrality. Denmark followed suit.'

Looking south, he said, 'Pressure on Greece in the early 'Seventies, which had resulted in a premature return to democracy, led later to a coup by the Communists.... Turkey, quick to realise her isolation, signed a friendship pact with the Soviet Union.'

Having continued with his doom-laden world tour, culminating with South-East Asia under Communism and Japan taking the place of the United States in Asian affairs, Walker turned to solutions. Stressing the urgency of strengthening NATO, he turned to his recurrent theme of morale. The people, he said, must be told the brutal truth about the power struggle. 'Politicians have been unable, or unwilling, to "sell" defence to the voters. If the voters were to be told the truth they would certainly talk to the politicians in the language they best understand – votes on election day.'

No further rebukes were received, perhaps because Walker was so soon to leave NATO and it was hoped that he would now be regarded as no more than a voluble eccentric. Before the end of February, 1972, the tiger would no longer prowl the NATO jungle.

The new Commander-in-Chief would make amends. General Pearson seemed heaven-sent. Handsome, charming and athletic, he was a diplomatist to his finger tips, as his current appointment as Military Secretary implied. He had been present at the liberation of Norway in 1945 and had married a Norwegian wife. Although he had not had Walker's experience of active service, he had long been an airborne warfare specialist and had commanded a parachute brigade. He radiated relaxed confidence and his air of calm command would soothe nerves if not summon up the blood.

The final weeks were filled with farewell visits, inspections and dinners. Cordiality prevailed even if sometimes tinged with relief. Fostervoll claimed that there had been no quarrel between them, just a misunderstanding. Yet he seemed to have learned much, for he showed real concern about the Russian acquisition of massive sea power and worried about threats of possible cuts in the Norwegian defence budget. Admiral Johannessen – himself about to retire – called and carefully avoided subjects of controversy. Privately, several senior Norwegian and Danish officers expressed gratitude and support.

Walker's final farewell visit was to Supreme Headquarters at Casteau. The planners there remained concerned about the northern flank: would Norway and Denmark join the European Economic Community with Britain and, if they did not, would they remain in NATO? If they remained, but Denmark persevered with disarmament plans, would other members of the Alliance – Belgium, for example – use this as an excuse to do likewise? What effect had Walker's attempts to galvanise the will to self-defence already had and what lasting effect would they have?

General Goodpaster gave a luncheon for the departing Commander-in-Chief, delivering a long speech of gratitude and one of the flowery diplomas that Americans love to present. As they left the table,

Goodpaster put a hand on Walker's shoulder and said, 'Well, we've had our differences, but in the end I think we learned to work together.'

Were the differences so great? Certainly there were differences in approach; far from 'playing it cool', Walker's first objective had been to 'make people sit up and take notice'. It was already being said in Whitehall that his appointment to AFNORTH had been a disaster and, although he was unrivalled as a field commander, he was not just a political simpleton but a dangerous meddler in politics. Walker himself always denied that he was a 'political general'. Yet was he such an innocent in the political jungle?

The essence of the Russian threat to Scandinavia was bullying. They tried to bully Norway and Denmark, as President Sukarno had tried to bully Malaysia. At sea, their naval exercises were pointedly aggressive. On land, their tank formations would charge up the Norwegian border and provocatively traverse their guns. Russian military aircraft swarmed on the NATO radar displays. Their propaganda proclaimed that NATO in Scandinavia was a threat to the Soviet Union and must be treated accordingly.

At last, somebody had stood up to the bully. An officer of the extinct Indian Army of the British Empire, a product of Quetta, Waziristan and Burma, a man left over from a past age. Walter Walker did not, at first sight, appear to belong to the time of inter-continental nuclear missiles, communications satellites, multimegaton hydrogen bombs and nuclear-powered submarines. Yet he had accepted and mastered the use of these weapons as naturally as he had once learned to handle a kukri. He did not seem cowed by the strength of the bully. He stood up and explained exactly how that bully could be faced and deterred.

On 7th February, Walter and Beryl Walker left their house at Holmenkollen and moved into an hotel in Oslo. Next afternoon, he and his successor discussed the inheritance and, at midnight, General Sir Walter Walker was succeeded by General Sir Tom Pearson as Commander-in-Chief, Allied Forces, Northern Europe.

Walter Walker retired from the British Army in the following May. Apart from the usual formalities there were no gestures of appreciation from his superiors and none can have been expected. Although offers of suitable employment had come from the Far East, these were refused because he felt that Beryl, after more than thirty years of camp-following, had more than earned a settled home. So, while preparing to continue making his views known, as a speaker and writer, Walker bought a seventeenth-century farmhouse in Somerset. As he began to restore the house and garden, his new neighbours were surprised to see the General attacking the overgrown shrubbery at the head of a party of Gurkhas with drawn kukris. The 1/7th Gurkhas, of which he was Colonel, were at Aldershot.

In September, 1972, Walker was approached by a somewhat timorous Ministry of Defence. The Indonesian Minister of Defence and National Security, on an official visit to London, had expressed a wish to meet him. Would it be possible for him to call at the Dorchester Hotel in London on the morning of 16th September to visit His Excellency General Maraden Panggabean?

So, less than a decade after they had faced each other in Borneo, the old adversaries met. General Panggabean greeted Walker warmly, saying that he had long been looking forward to the occasion. They spoke of their campaign and the Indonesian congratulated the Englishman on the quality of his intelligence in Borneo. Walker told him about the work of the Special Air Service and the Border Scouts and stressed the importance he had attached to winning the hearts and minds of the people.

Panggabean said that he had recently been in Malaysian Borneo on an official visit and the two men discussed the problems of the territory for which they had fought. Before the meeting ended in friendly accord, Panggabean handed Walker a parcel.

When he returned to the United Service Club for lunch, he unwrapped a carved wooden figure, an example of the arts that Indonesia share with Malaysia. This now stands in the Somerset farmhouse. More than any official honour or award, more than any commendation, that wooden figure represents a triumph. Acclaim from an old enemy can sometimes be of wider and deeper value than the acclaim of friends.

About the Author

Tom Pocock is the author of 18 books (and editor of two more), mostly biographies but including two about his experiences as a newspaper war correspondent.

Born in London in 1925 - the son of the novelist and educationist Guy Pocock - he was educated at Westminster School and Cheltenham College, joining the Royal Navy in 1943. He was at sea during the invasion of Normandy and, having suffered from ill-health, returned to civilian life and in 1945 became a war correspondent at the age of 19, the youngest of the Second World War.

After four years wth the Hulton Press current affairs magazine group, he moved to the Daily Mail as feature-writer and then Naval Correspondent, becoming Naval Correspondent of The Times in 1952. In 1956, he was a foreign corresponent and special writer for the Daily Express and from 1959 was on the staff of the Evening Standard, as feature writer, Defence Correspondent and war correspondent. For the last decade of his time on the Standard he was Travel Editor.

He wrote his first book, NELSON AND HIS WORLD in 1967 on his return from reporting the violence in Aden and his interest in Nelson has continued. Indeed, eight of his books are about the admiral and his contemporaries; his HORATIO NELSON was runner-up for the Whitbread Biography Award of 1987.

Tom Pocock has contributed to many magazines and appeared on television documentaries about Nelson and the subject of another of his biographies, the novelist and imperialist Sir Rider Haggard.

[1] This is the Army's style of referring to the 1st Battalion of the 8th Gurkha Rifles.

[1] *A Child at Arms* (Hutchinson, 1970).

Printed in Great Britain
by Amazon